顶 级 设 计 空 间
TOP DESIGN SPACE

主编 张青萍 · id+c《室内设计与装修》杂志社

奢华酒店
LUXURY
HOTEL

五星级酒店 & 艺术酒店
FIVE-STAR HOTELS
ART HOTELS

（第二版）

中国林业出版社

图书在版编目（CIP）数据

奢华酒店：汉英对照／《顶级设计空间》编委会编
. -- 2版. -- 北京：中国林业出版社, 2013
　（顶级设计空间）
　ISBN 978-7-5038-7275-4

Ⅰ.①奢… Ⅱ.①顶… Ⅲ.①饭店－室内装饰设计－
图集 Ⅳ.①TU247.4-64

中国版本图书馆CIP数据核字(2013)第274860号

《 顶 级 设 计 空 间 》 编 委 会 编

主编：张青萍
编委：孔新民、贾陈陈、许科、李钢、吴韵、竺智、曾丽娴

责任编辑：纪亮
英文翻译：董君、梅建平、牛晓霆、万毅、赵强

出版：中国林业出版社
　　　（100009 北京西城区德内大街刘海胡同 7 号）
网址：www.cfph.com.cn
E-mail：cfphz@public.bta.net.cn
电话：（010）83143581
发行：新华书店
印刷：北京卡乐富印刷有限公司
版次：2016年4月第2版
印次：2016年4月第1次
开本：230mm×300mm
印张：32
字数：360千字
定价：199.00元

设计进行时

中国改革开放30年，室内设计行业行进到今天，也已经有了无计其数的变化、发展和积累，30年的思考、30年的实践、30年的进步，也造就了这30年的成绩。

我们好似在进行着一场接力赛，祖先把中华民族灿烂的文化一代代地传承到21纪，我们有责任将这份优秀的遗产倍加珍藏以传给后代。我们所面临的挑战是拿什么当代的室内文化馈赠后人？但其实这30年中由于电子技术的普及和信息的迅速传播和交换，设计已出现国际化、同一化的倾向，与此同时引起了传统性、地域性和个性差异的不断丧失，又由于社会追求物质与功能价值的同时造成对精神和文化价值的忽视，我们已找不到回头的路。但不管历史结论会如何，我们这代人是努力的、勤奋的，是不断地用自己的智慧为中国室内设计行业进步奉献着的。

总的看来，21世纪的室内设计发展有以下倾向和趋势：

倡导绿色设计

人类起源于自然，其间虽曾摆脱过自然，但最终还是要以全新的面目去回归于自然。如此轮回恰恰历史地、辩证地道出了人与自然关系的变化。如今的人们，特别是生活在大城市里的人离大自然已越来越远了，于是人们特别希望在室内再现一些大自然的情景，以求得哪怕是暂时地、局部地享受。作为设计师一方面尽可能地创造出生态环境，让人们最大限度地接近自然，另一方面须有环保意识，努力去提高设计中的健康因素，以满足人们在生理和心理上的需要。

室内设计中的健康设计充分利用自然或仿自然的因素，为人们提供生活舒适的空间。室内的色彩、照明及功能空间的弹性分割，都应该在满足其基本功能的基础上，尽可能充分利用自然能源。尤其是提倡对绿色装饰材料和绿色照明材料的运用，同时注重社会心理学的研究。绿色设计本着以下几个原则：

①设计上使用最少的材料，少浪费应节约的资源、能源，力戒奢靡；②尽量多采用污染少，环保性能强，安全可靠的材料；③符合人体工程学的要求，讲求空间上比例与尺度，避免使人感到压抑或繁琐；④设计应以人为本，满足人的本质需要。

运用高科技手段

科技进步影响着人类生活的方方面面，现代科技的发展为人类的衣食住行提供了很多方便，它可以使居住环境更加符合人们的意愿。因此，在设计中提高科技含量，创造高效率、高功能、高质量室内生活环境的要求已愈加鲜明。

在本世纪，楼宇的智能化将逐步实现。建筑智能化就是将结构、设备、服务运营及相互关系进行全面综合配置，从而达到最佳的组合程度，使建筑具有高效率、高功能和舒适性。单从自动化来看，就是实现建筑设备自动化、办公自动化和通讯自动化。建筑智能化并非仅仅针对大楼宇，还会迅速走向寻常百姓家。住宅中自动防盗报警，自动调温、调湿、自动除尘、调节灯光亮度、自动控制炊事用具等，如今也已成为事实。

注重设计的系统性

在这个日新月异、急剧变化的时代，系统地看待问题和解决问题是当代人的特质。站在现代与历史之间的人们既希望从传统中找回精神的家园，以弥补快速发展带来的心理失落与不安；同时又满怀着激情和野心试图运用当代技术和审美重新诠释历史，使之适应现代生活。系统的眼光可使我们将面临的室内设计浅层次问题渗透到更深的层次方面加以科学、综合地解决。这样的设计系统性包括环境学、生态学、经济学、系统论、方法论、控制论、统筹学、管理学及有关室内设计方面的政策法规、标准规范等方面的内容。

室内设计系统是指应用系统的观点和方法，将室内设计的内容、要素，相关的领域和环节，以及室内设计的程序予以统筹而形成的一个框架体系。从与其相关部分的关系和进行的程序来分析，可理解为有横向设计系统和纵向设计系统两个方面。横向系统设计表现为在设计过程中所涉及到的如生理学、心理学、行为科学、人体工程学、材料学、声学、光学、经济学等诸多因素；纵向系统设计表现为对设计实现过程中所有历程的考虑。概括而言，横向系统设计强调相关与联系；纵向系统设计强调过程与变化。

无论将来的室内设计如何发展，它都必然在一个更光泛、更全面的系统里科学地伸展，设计师也必将持有这种科学的态度和掌握这类理性的方法而从事设计。

本系列图书刊登的是近年来一些顶级的设计作品，它们或多或少的反映的是当下室内设计师的思考和当前技术背景下的实践。30年告了一个段落，下一个30年又将开始，我们已走上新的征程，设计永远是进行时。愿这本套书的出版能得到业界的认可和赞扬！

南京林业大学 风景园林学院副院长、教授
id+c《室内设计与装修》杂志主编 张青萍
2010.3.1草于南京

五星级酒店

艺术酒店

Five-Star Hotels

Art Hotels

FIVE-
HOT

五星级酒店

STAR HOTELS

HILTON
TM IN BEIJING
北京希尔顿逸林酒店

01

【坐落地点】北京广安门外大街168号；【面积】4000 m²
【设计】梁景华
【室内设计】香港PAL设计事务所有限公司
【摄影】鲍世望

落户北京的亚洲首家希尔顿逸林酒店（Doubletree by HiltonTM）旗舰店，建筑设计匠心独具，将东方哲学思想与国际艺术设计完美融合。外观设计以"水滴"为核心概念，凸显水之灵动与自然。而在内部装饰上，北京希尔顿逸林酒店的所有艺术作品均由世界各国著名艺术家共同打造，与店内整体设计风格相得益彰，将酒店内的艺术氛围烘托得无以复加，为客人带来高品质享受。

酒店的10个会议室和宴会厅巧妙地围绕空中花园而建，形成了一个匠心独具的整合会议功能区域。设计师将园林风情中提炼出的自然神韵运用到城市户外花园的空间布局中，各景观自成一体又相映成趣，整体层次清晰，高低错落而又交相辉映。

所有陈列于酒店的艺术品可以在大堂、餐厅、会议室、电梯、健身中心及行政楼层欣赏到。风格高雅、现代抽象，却能雅俗共赏，与店内整体设计格调相得益彰，将酒店的艺术氛围烘托得美轮美奂，宛如一场精心准备的艺术饕餮盛宴。

As Asia's first Doubletree by HiltonTM flagship store, architectural design creative design, the oriental philosophy and international art and design the perfect fusion. Designs with "drop" as the core concept, highlighting the smart water and nature. In the interior, the hotel's all the works of art by famous artists around the world work together to build, and store the overall design style complement each other, will heighten the artistic atmosphere of the hotel was to be added, in order to bring high-quality enjoyment of guests.

The hotel's 10 meeting rooms and banquet hall cleverly built around the sky gardens, forming an integrated conferencing capabilities creative design of the region. Designers will be extracted out of style gardens of natural charm to the city the use of outdoor garden space layout, each self-contained landscape also exist side by side, the overall level of clear, but echo the high and low scattered.

All works of art on display in the hotel in the lobby, restaurant, conference rooms, elevators, fitness center and executive floors to enjoy. Style, elegance, modern abstract, but it can widely appealing, and store the overall design style complement each other and contrast the artistic atmosphere of the hotel was beautiful, like a well-prepared art gluttonous feast.

↑↑红色的巨型艺术品在沉稳的空间极为出挑 / The red giant works of art in the calm of the space is extremely significant.
←大堂接待台 / Lobby reception desk.
↓一层平面图 / 1st floor plan.

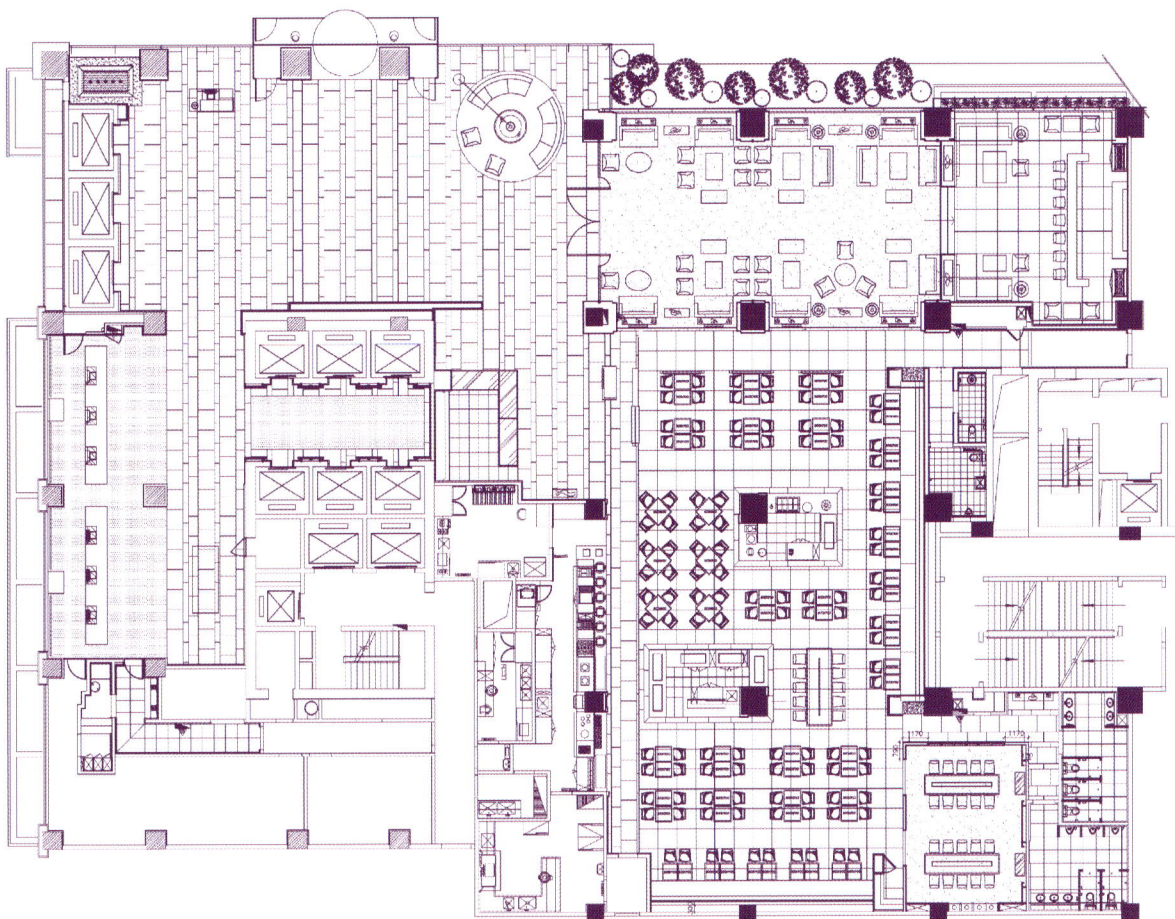

↑↑↑粗犷的艺术品 / Rough work of art.
←走廊上的休息区 / Corridor on the rest area.

←电梯厅的艺术品，展现现代东方的哲学思想 / Elevator hall of art, displaying the modern Eastern philosophy.
↑随园中餐厅入口处 / Chinese restaurant at the entrance.

↑ VIP包间休息区 / VIP rooms rest area.
← 中餐厅 / Chinese restaurant.
↓ 大堂吧 / Lobby Bar.

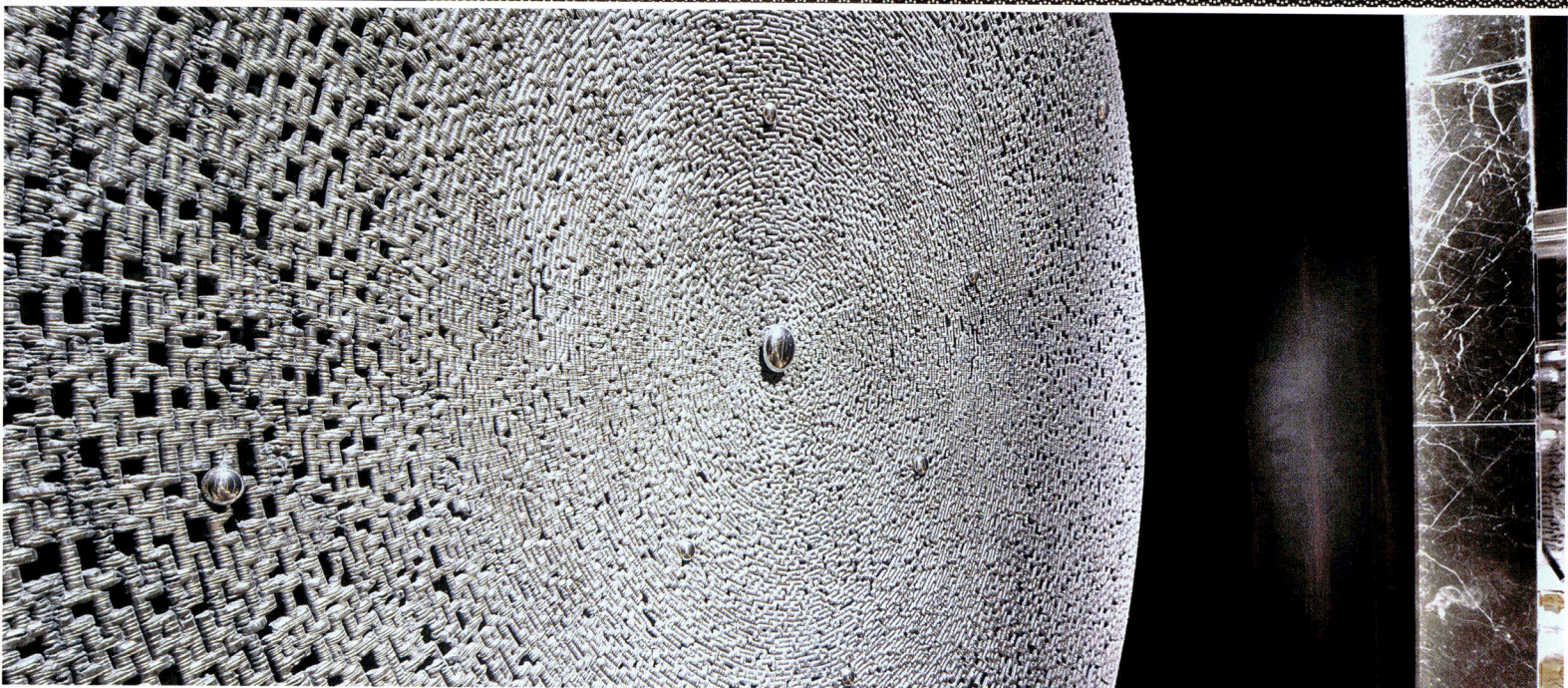

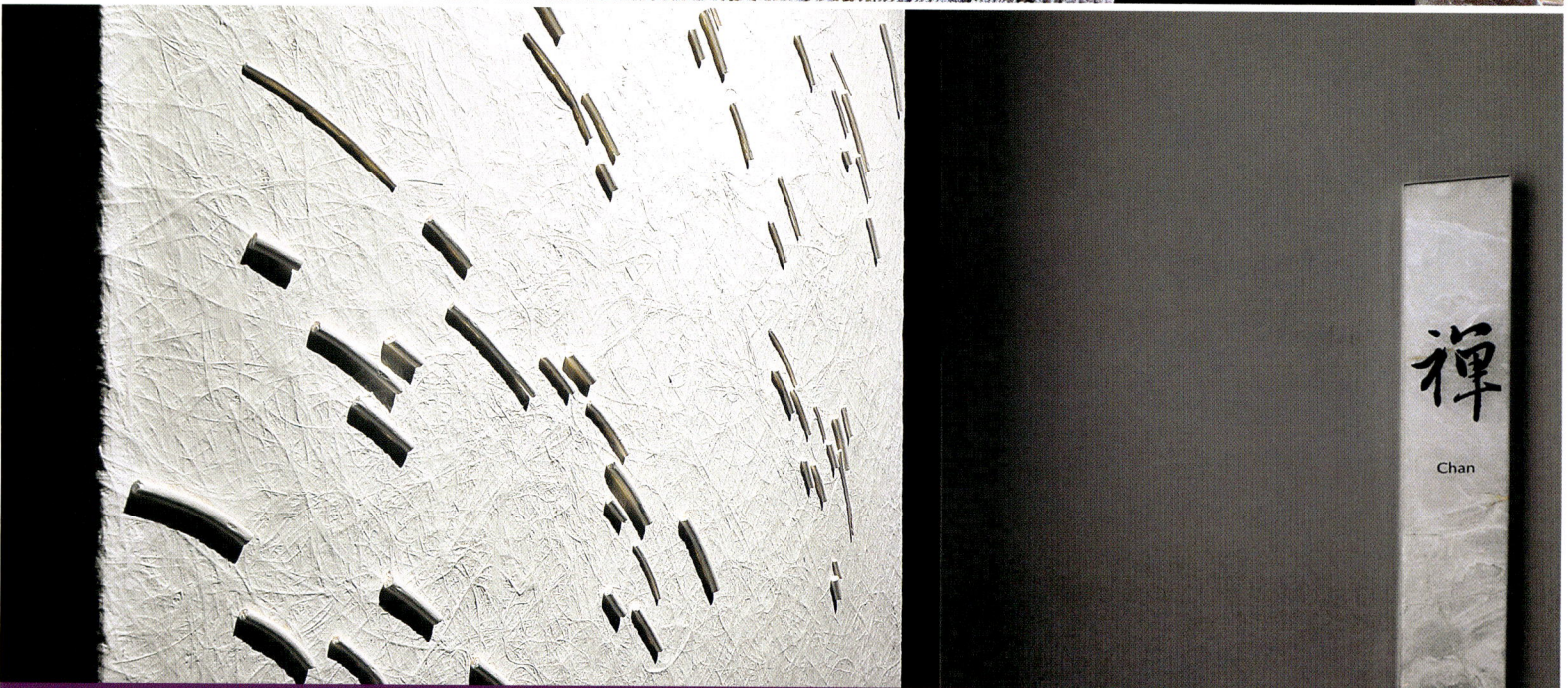

↑墙面上的艺术品，展现现代东方的哲学思想 / Works of art on the wall, displaying the modern Eastern philosophy.

JW MARRIOTT
HOTEL

金茂深圳JW万豪酒店

02

【工程名称】金茂深圳 JW 万豪酒店
【面积】37 611 m²
【坐落地点】深圳福田区
【主要材料】横纹玻璃、白色　油金属框架、白云石

万豪酒店位于高级写字楼林立的深圳市中心城区，近邻天安数码城。区位决定客源，JW万豪酒店必然是将现代商务人士定为目标客户。大气多元应该是业主方所想要的效果，整体风格与使用的设计元素相对时尚，极具现代感，同时内敛沉稳的商务特征明显。

原先的商务楼构造，使得内部有大量的立柱、电梯，同时空间的分割亦与酒店布局不相吻合。这就给设计师带来了前所未有的挑战，同时激发了他们大胆运用建筑效果和基调创造出令人惊艳的空间。穿过刻意设计的"窄小"大堂前厅，视野豁然开朗，配合二层较高的中庭，达到了移步换景的视觉效果。

二楼的餐厅区包含有中餐、西餐、日本餐等各式餐饮。中餐区不是很大，置于面向公共走廊上的开放式厨房，打破老式酒店厨房设计的死板，让空间更加灵动。值得一提的是这里地道的瑶柱粥为整个餐厅增色不少，让人留恋。日本料理区的原生石材缓和了酒店太商务、太硬朗的氛围，服务台具有连贯性。宴会厅设计得比较柔和，运用了中国剪纸手法的设计元素，简约却不失细节的表达，使得空间设计饱满而文雅。

SPA区的设计是酒店中较有特色的一部分。前厅的装饰设计灵感来自于"水泡"，氛围温和清新。围合成圆形的休息厅，具有极佳的视觉效果。

JW Marriott Hotel is located in Shenzhen high-level office buildings of downtown, close neighbors, Tian´an Cyber Park. Location decision of tourists, the hotel would certainly be a modern business people as target customers. The atmosphere should be the owners of multi-party by the desired effect, the overall style and design elements used in the relative fashion, very modern at the same time restrained obvious characteristics of steady business.

The original commercial buildings constructed, making a large number of columns inside, elevators, while the division of space with the hotel does not correspond to the layout. This gives designers unprecedented challenges, and inspire their bold use of architectural effects and tone to create a stunning space. Through the deliberate design of "narrow" lobby lobby, vision suddenly see the light, with the two-story high atrium, reaching the venue for King of the visual effects.

On the second floor of the dining area contains a Chinese, Western, Japanese dishes and other kinds of food. Chinese food area is not large, placed in the corridor for the public open kitchen, breaking the old hotel kitchen design rigid to allow space for more Smart. It is worth mentioning here was typical of the dried scallop congee being groomed for the entire restaurant, people yearn. Japanese Restaurant Area native stone to ease the hotel is too commercial, too tough atmosphere, the service desk coherence. Banquet hall designed to be softer, using the Chinese paper-cutting practices of design elements, the expression of simplicity but without losing the details, making full and elegant space design.

SPA area is designed to be a hotel as part of the more distinctive. Lobby decoration design inspiration from the "bubble", a mild fresh atmosphere. Circular lounge, with excellent visual effects.

←↑大堂中庭的光柱有一种宏大的气势 / Beam atrium lobby has a grand momentum.
↓二层平面图 / 2nd floor plan.

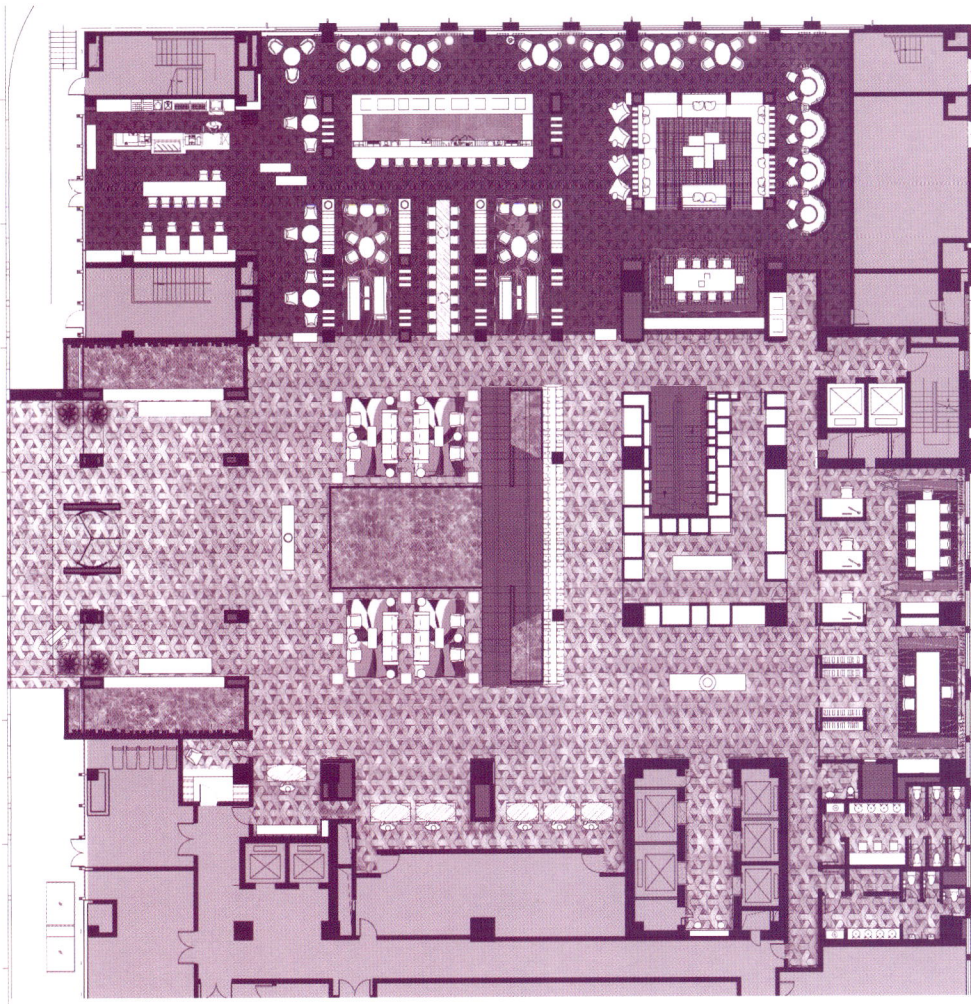

↑中式餐厅区的茶壶摆设 / Chinese restaurant area teapot ornaments.
→中式餐厅区极富风雅 / Chinese restaurant area very elegance.
↓餐厅区一角 / Dining area corner.

↑ "泉" Spa的前厅设计感十足 / "Quan" Spa in the lobby of the design sense of the full.
↓Spa的设计理念来自喷涌而出的泉水 / Spa design concepts from spewing out of the spring.

↑客房内的豪华设施和明亮色彩 / The luxurious room facilities and bright colors.
↓客房平面图 / Room plan.

FONTANA PARK
HOTEL
Fontana Park 酒店

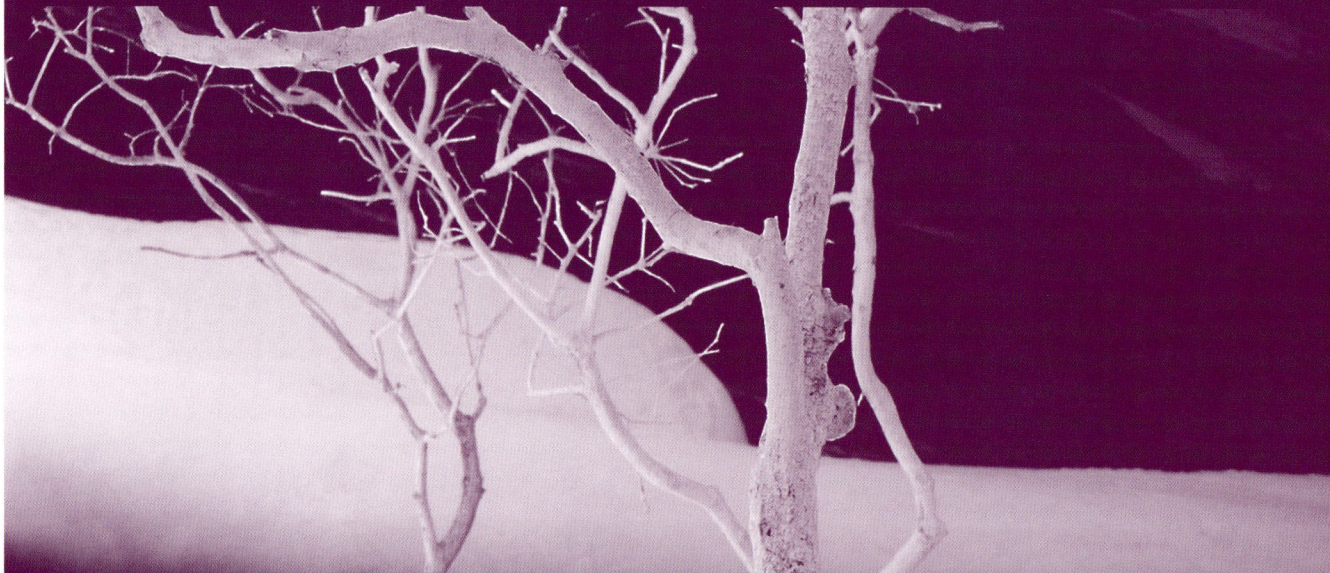

03

【坐落地点】葡萄牙里斯本
【设计】Nini Andrade Silva 〔葡〕
【建筑设计】Francisco Aires Mateus
【摄影】Coutesy of Nini Andrade Da Silva

酒店位于里斯本的中心地带，以极佳的地理位置和风格化的时尚设计而闻名。酒店既接近历史文化地区、商业中心，又临近海滩和机场，便捷的交通区位，成为游客的上佳选择，同时，时尚的设计酒店风格又将它与其他酒店区别开来。在喧闹的市中心，用细腻的设计手法展现现代葡萄牙的风貌，同时颇有匠心地将东方"禅意"融入空间设计，为四海旅人提供静谧休息的绝佳处所。

极简艺术的自如运用，弱化了空间的功能性，将Fontana Park酒店打造得更像一个偏重概念的展厅，置身其中全无里斯本市中心的喧闹浮躁之感，更多是后现代的冷静克制。大厅中铁架下的黑色蒲团和低矮的地桌，散着淡淡的东方"禅意"。整个酒店通过很多这种不刻意的设计，将东方的"禅意"隐没在极简设计主题的背后。东方文化的内敛配合后现代的色彩搭配，而突显了"禅意"的亘古不衰。

为忠实建筑原本的钢架结构，设计师仅通过改变钢铁的外观色彩和造型，来软化铁给人的坚硬冰冷之感。接待台上空的白色电线，以舞动的姿态出现在旅客面前，这种刻意而为的混乱反倒使得它们极具艺术张力，一改电线聚集时给人的杂乱无序感。同时，设计师还通过木质材料变换运用，来左右整个空间的氛围。

Hotel is located in the heart of Lisbon, to excellent location and style of fashion design is known. Hotels close to both the historical and cultural district, commercial center, close to the beach and the airport, and convenient transportation location, a good choice for tourists, while the stylish design hotel style, turn it with other hotels to distinguish. In the bustling city center, with delicate design technique to show the modern Portuguese style, while quite ingenuity to the mood of the East into the space design for the whole world travelers the perfect premises to provide quiet rest.

Ease the use of minimalist art, weakening the functionality of space, will be more like Fontana Park Hotel to create an emphasis on the concept of exhibition, exposure to which no hustle and bustle of downtown Lisbon, impetuous feeling more calm and restraint in the post-modern . Hall of formwork under the futon and the low ground black tables, the Oriental exudes mood. The entire hotel through many of these are not intentionally designed into the mood of the East behind the minimalist design theme. Restrained oriental culture with the post-modern mix of color and highlights the eternal Oriental mood.

Steel frame construction for being faithful to the original structure, designers just by changing the appearance of color and shape of steel, to soften the hardness of iron gives the feeling of cold. Reception table empty white wire to dance would appear on the passenger front, this deliberate confusion of hand, but for making them extremely artistic tension, a change wire disorderly aggregation gives a sense of. At the same time, designers transformation through the use of wood materials to the atmosphere around the whole space.

↑大厅接待台 / Lobby Reception Desk.

↑大堂的"禅意"一角 / Hall's "Zen" corner.
↓休闲吧 / Lounge

↑休闲吧 / Lounge
↓平面图 / Plan

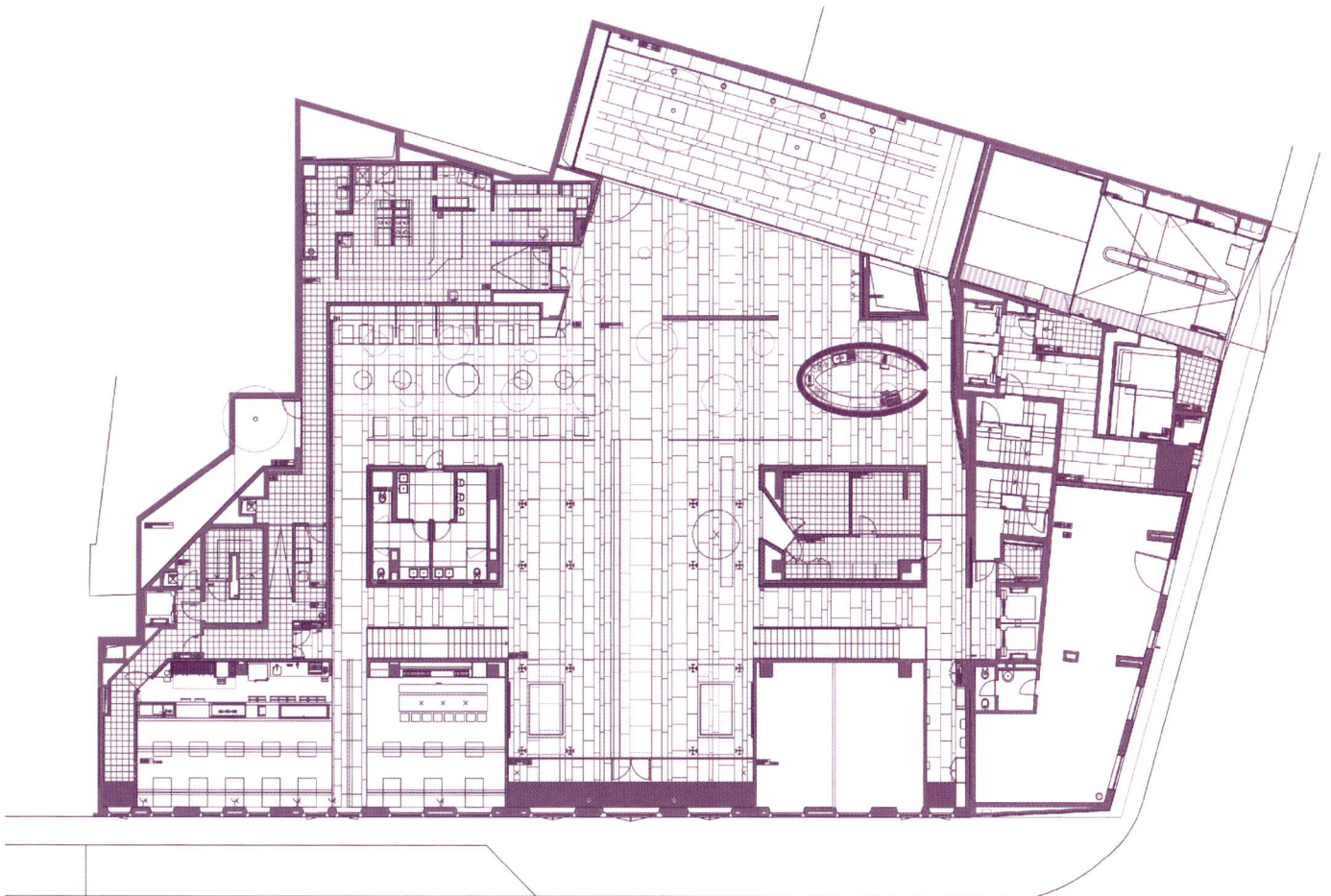

↑Saldanha Mar餐厅的纯粹让人难忘 / Saldanha Mar Restaurant purely unforgettable.
→Fontana酒吧一角 / Fontana Bar corner.
↓Saldanha Mar餐厅一角 / Saldanha Mar Restaurant corner.

↑Fontana酒吧过道 / Fontana Bar aisle；↑客房外的过道 / Hallway outside the room.
↓Fontana酒吧一角的藤椅 / Fontana Bar corner wicker chair；↓Fontana酒吧别致的地灯 / Fontana Bar chic lamp.

↑标准间 / Standard room.
↓↓标准间的俯视图 / Standard between the vertical view.

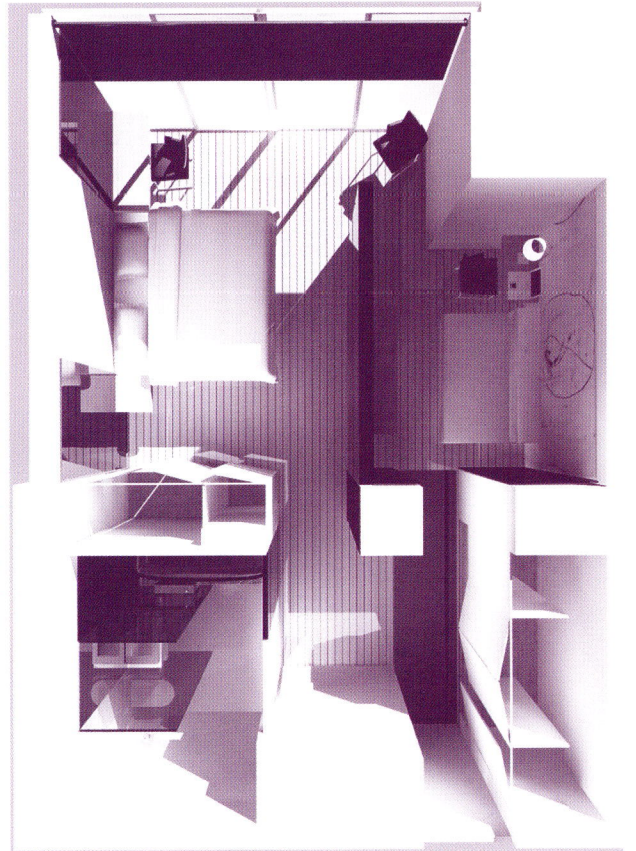

OPPOSITE
HOUSE

瑜舍酒店

04

【坐落地点】北京三里屯11号院1号楼
【面积】12 000 m²
【设计】隈研吾 NHDRO设计事务所
【摄影】舒赫

瑜舍是一个豪华时尚酒店，它是香港太古集团"三里屯Village"的一部分内容，为低密度综合商业项目。据说"瑜舍"得名于一个中国汉字，指院落中正房对面的建筑，通常为房主的贵宾所居，看来这又是一个深藏东方传统文化内涵的建筑。东方传统居住文化的表达总是难以离开我们既亲切又为之骄傲的院落，因为在我们对传统的记忆中院落是精神和物质的核心，也是文化活动的空间载体。

瑜舍一共设有99间客房，分6层，环绕于大堂的四周。99间客房之中半数以上的面积超过了70 m²，房间之大北京罕有，其中45 m²的有43间，70 m²的有46间，95 m²的有7间，还有面积达115 m²的开放式客房两间，以及180 m²顶层复式带空中院落的套间一间。客居者和院落的关系乃是设计者努力刻画和表现的主与客之间的相互关联的一种物质表达。

客房环廊的处理亦是公共区域设计之中的亮点，走廊的一侧是极简约风格的玻璃栏板，精致的构造处理使栏板保持了最大程度的通透状态，向中央那个巨大的共享空间敞开着。走进这条走廊，立时一种依稀朦胧的乡土气息扑面而来，将人诱入时间的陷阱之中。同时这种连续性和整体性的强化也达到了一种隐蔽的效果，它强有力地创造出一个界面来划分共享和私人的空间，不能否认这是一种对个人和公共性质之间的"有趣"平衡。

Opposite House is a stylish luxury hotel, which is Swire Group of Hong Kong, "Sanlitun Village" part of the content for low-density integrated commercial project. Is said to "Opposite House" named after a Chinese character that courtyard Chiang Kai-shek room opposite the building, usually the owner of the guests of the home, it seems this is also a deep oriental traditional culture of architecture. The expression of the traditional oriental culture is always difficult to live away from us both a cordial and also proud of the compound, because in our memory of the traditional courtyard in the heart of the spiritual and material, but also spatial carrier of cultural activities.

Opposite House has a total of 99 rooms divided into six layers, surrounded by four weeks in the lobby. 99 rooms among the more than half the size of more than 70 m², a room big Beijing rare, of which 45 m² are 43, 70 m² there are 46, 95 m² of 7, as well as an area of 115 m² of open-ended Room, as well as 180 m² penthouse suite with an air courtyard. Staying as a guest and the compound to characterize the relationship between the designer and performance Naishi primary and a van between the expression of a substance related.

Room corridor treatment is the focus of the design of public areas, corridors on one side is a very simple style of glass tailgate, fine structure treatment to tailgate with the utmost degree of permeability state, to the central shared space for that great wide open . Into the corridor, immediately a faint hazy country flavor rushing toward us, which will lure into the time trap. At the same time the continuity and integrity of the reinforced also reached a hidden effect, it is a powerful way to create an interface to divide shared and private space, can not deny that this is a right between the personal and public character " an interesting "balance.

↑大厅既是酒店的大堂，也是一个当代艺术的展厅 / Hall is the hotel's lobby, but also a contemporary art exhibition hall.

037

↑ "瑜舍" 的大厅 / "Opposite House" of the hall.
↓ 大厅的艺术墙 / Art hall walls.

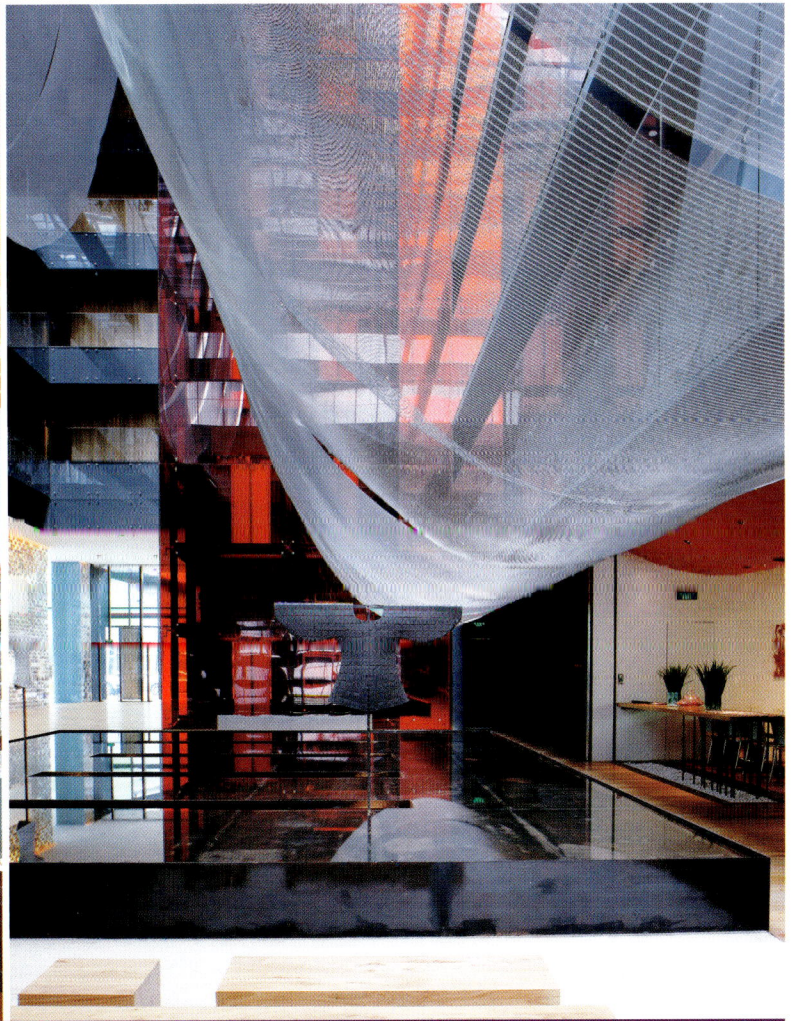

↑↑ "瑜舍"大厅的艺术展示空间 / "Opposite House" lobby art exhibition space.
↓大厅的全貌 / Complete picture of the hall.

039

↑餐厅入口 / Restaurant entrance.
→Sureno餐厅的休息区 / Sureno restaurant's lounge area.
↓Sureno餐厅 / Sureno restaurant.

↑Sureno餐厅的艺术灯 / Sureno restaurant art of light.
↓Bei餐厅入口 / Bei restaurant entrance.

↑ Bei餐厅 / Bei Restaurant.
↓ 餐厅中绚丽的灯光 / Restaurant in the magnificent light.

↑ 走廊的设计 / Corridor design；↑ 大面积使用了木材的走廊 / Corridor on the use of large timber.
←客房环廊的设计是公共区域中的亮点 / Room corridor is designed to be a bright ideas in public area.
↓大面积使用了木材的走廊 / Corridor on the use of large timber.

↑客房 / Rooms
↓换个角度看客房 / Rooms from another angle.

↑ 客房的局部处理 / The partial treatment rooms.
↓ 木制的浴室 / Wooden bathroom.

TRIANON
PALACE HOTEL
Trianon Palace 酒店

05

【坐落地点】法国巴黎
【规模】2幢宫殿式建筑，199间客房，2800 m²温泉区
【设计】JRichmond International 〔英〕
【摄影】Courtesy of Richmond International

新的接待处，黑色的接待台稳重大方，墙面的深灰色给人如家般的温馨感受，手工的小地毯色彩鲜亮。而墙面上最引人注目的是一个硕大的表盘，表盘的设计灵感来自于卡迪亚手表的经典样式。走廊的地面采用了几何形状拼贴大理石，整齐庄重，这也是古典主义常用的设计手法。而现代的家具样式让人很明显地体会到酒店设计的层次感。

宫殿里的每一个客房都有各自不同的布局：在现有的房间基础上用一些明快的颜色和图案勾画出房间的轻快感觉，而在细节方面同样仍然采用了传统设计里面复杂的工艺和造型。由于受到城堡装饰风格的启发，酒店里所有的家具都带有贵族气质。这些定制的家具无一例外使用了豪华的材料。

设计师结合建筑原有的结构把餐厅处理成3个紧密相连又可以独立就餐的区域，它们精巧而不失华丽，优雅而舒畅。建筑大尺度的层高给餐厅带来了很好的空间感，而相对应比例的落地窗户显示了皇家特有的气派。大面积落地镜面的酒水柜是受到法式服装店设计的影响，通透华丽。而装饰柜通体印有一幅巨大的凡尔赛宫的照片，在柔和的光线下自然地呈现。

The new reception area, a black stable generous reception station, the dark gray walls give a warm, such as home-like feel, handmade rugs bright colors. As for the walls of the most striking is a huge dial, dial design inspiration from the classic style of Cartier watches. The corridor on the ground marble with a collage of geometric shapes, clean-minded, which is commonly used in classical design techniques. And modern style furniture makes it clear that the level of experience to the hotel a sense of design.

Each of the palace rooms have different layouts: on the basis of the existing room with some bright colors and patterns outlined lively sense of the room, while the details are still using the same traditional design inside the complex processes and forms . As a result of decorative style inspired by the castle, the hotel where all the furniture with an aristocratic temperament. Without exception, these custom furniture using luxurious materials.

Designer with the architectural structure of the original restaurant is processed into three closely linked and can separate dining area, they are exquisite to the public yet beautiful, elegant and comfortable. Large scale-storey building to the restaurant to bring a good sense of space, while the corresponding proportion of floor window shows the royal unique style. Large floor mirror drinks cabinet is subject to French clothing store design, by permeability gorgeous. The cabinet henselae printed a large photograph of the Palace of Versailles, in the soft rays of the sun naturally present.

↑ 酒店接待处 / Hotel reception.
→↓ 各自独立又能贯穿起来的餐厅 / Independent but also runs through the restaurant together.

↑贵宾厅，超大的空间显示出皇家礼遇 / VIP room, large room show royal courtesy.
→↓啤酒吧的配套餐厅，有浓郁的宫廷风格 / Beer Bar supporting restaurants, a rich palace style.

↑ 简洁明快的各式客房 / Concise variety of guest rooms.
↓ 简洁明快的各式客房 / Concise variety of guest rooms.

055

↑ 豪华套房的会客厅 / Deluxe suite living room.
↓ 豪华套房的餐厅 / Luxury suites conference restaurant.

↑ 豪华套房的客房 / Deluxe suite rooms.
↓ 豪华套房的卫生间 / Deluxe suite bathroom.

PARK HYATT
SHANGHAI

上海柏悦酒店

06

【坐落地点】上海浦东世纪大道
【面积】1200 m²；【设计】季裕棠设计事务所 〔美〕
【设计公司】公共区域，上海康业建筑装饰工程有限公司；客房区域，香港海华设计事务所
【摄影】贾方

整个酒店通过素淡雅致的色调和藏而不露的现代艺术品设置，完美诠释一座现代顶级精品酒店的高雅华贵。酒店公共区域分别位于环球金融中心的底层及85、86、87层，包括酒店入口、游泳池、接待、茶吧、酒吧、西餐厅、酒廊、会议室等功能区。174间中式客房，为住客打造沉静的私享空间。

柏悦酒店一改以往酒店金色、米黄色系的传统手法，而以灰色、白色、咖啡色为主基调，意求低调的奢华，简约的华丽，色系效果素雅和谐，并将富有创造力的艺术饰品贯穿其中，营造出一种中国水墨画的写意。

在满足功能要求的情况下，空间如何合理安排是设计的重点难点。酒店的接待台位于高低电梯转换的通道处，并利用四组内置艺术品的家具作为接待区的背景，形成一个半围合区域，也是一种新的手法，处理上并没有特意把此功能区做过多地渲染，但与茶吧的有机结合，把窗外最好的风景留给了客人，真正做到了功能与服务的完美结合。

从灰色的木皮、大理石、水晶墙、吊顶、白铜金属、鱼皮、羊皮、钢琴漆、银箔艺术玻璃、幕墙卷帘到大量的定制灯具，都是为了营造现代中式空间而度身打造的，新材料的运用往往会产生一些意想不到的效果。

Through the entire hotel tones and modern art decoration, the perfect interpretation of a modern top-level luxury boutique hotel elegance. Hotel public areas are located at the bottom of a global financial center and the 85,86,87 layer, including the hotel entrance, swimming pool, reception, tea bar, bars, restaurants, lounges, meeting rooms and other functional areas. 174 Chinese-style rooms for the residents to build quiet private space to enjoy.

Hotels have changed their hotel gold, beige color of the traditional methods, while the gray, white, brown the main tone, intended to seek low-key luxury, simplicity of the gorgeous color effect is simple and elegant and harmonious, and creative art of jewelry through the Among them, creating a sense of Chinese ink painting freehand.

In meeting the functional requirements of the situation, how space is a reasonable arrangement designed to focus on difficult points. Hotel Reception Desk is located in the channel at high and low lift conversion and take advantage of four built-in furniture, works of art in the background as a reception area, forming a semi-enclosure area, is also a new approach to treatment is not specifically for this functional area to do too much exaggerated, but the organic tea bar combination of the best scenery of the window left to the guests and truly the perfect combination of features and services.

From the gray veneer, marble, crystal, ceiling, metal, skin, sheepskin, piano, art glass, silver foil, rolling to the wall a large number of custom lamps, are designed to create a modern Chinese space tailor, and new materials use tends to produce some unexpected results.

↑ 夜色中的上海浦东环球金融中心 / Night of the Shanghai World Financial Center in Pudong.

↑挑高的天井 / High-ceilinged courtyard,　↑大堂的接待处 / Lobby reception,
←酒店位于上海浦东环球金融中心79至93层 / Hotel is located in Pudong, Shanghai World Financial Center, 79-93 layers.
↓走廊的当代艺术品 / Corridors of contemporary art.

↑楼梯间的当代艺术品 / Stairwell contemporary art.
↓87层平面图 / 87th floor plan.

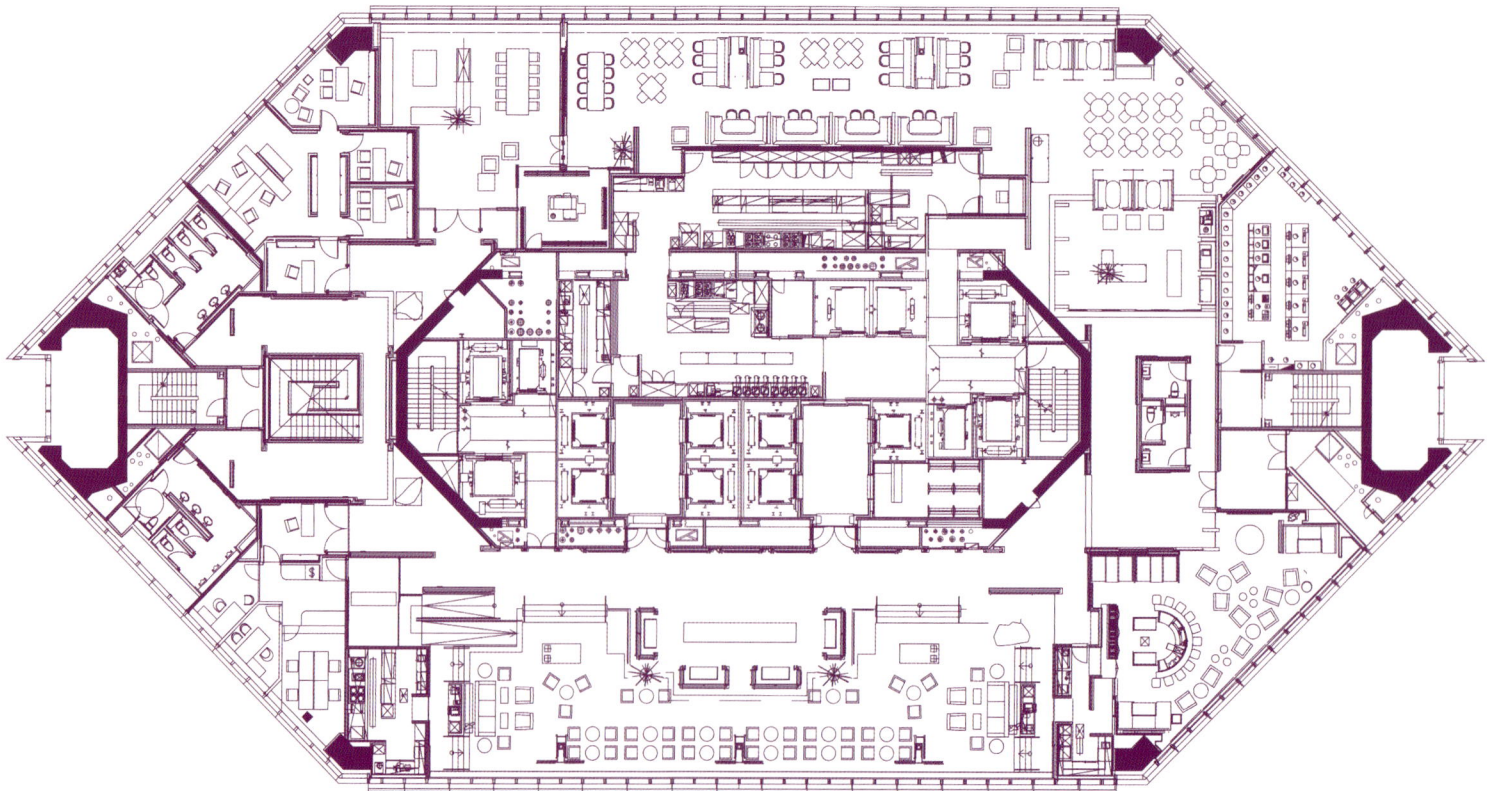

↑87层的餐厅入口 / 87th floor restaurant entrance.
↓87层的餐厅的外隔断 / 87th floor restaurant outside the walls.

↑↑87层的开放式餐厅 / 87th floor, an open restaurant.
→包间采用纯欧式的装饰 / European-style rooms decorated with pure.
↓ 87层的封闭式品酒间 / 87th floor, closed-end tasting rooms.

↑84层的客房卧室 / 84th floor bedroom rooms.
←84层的客房一角 / 84th floor bedroom corner.
↓84层平面图 / 84th floor plan.

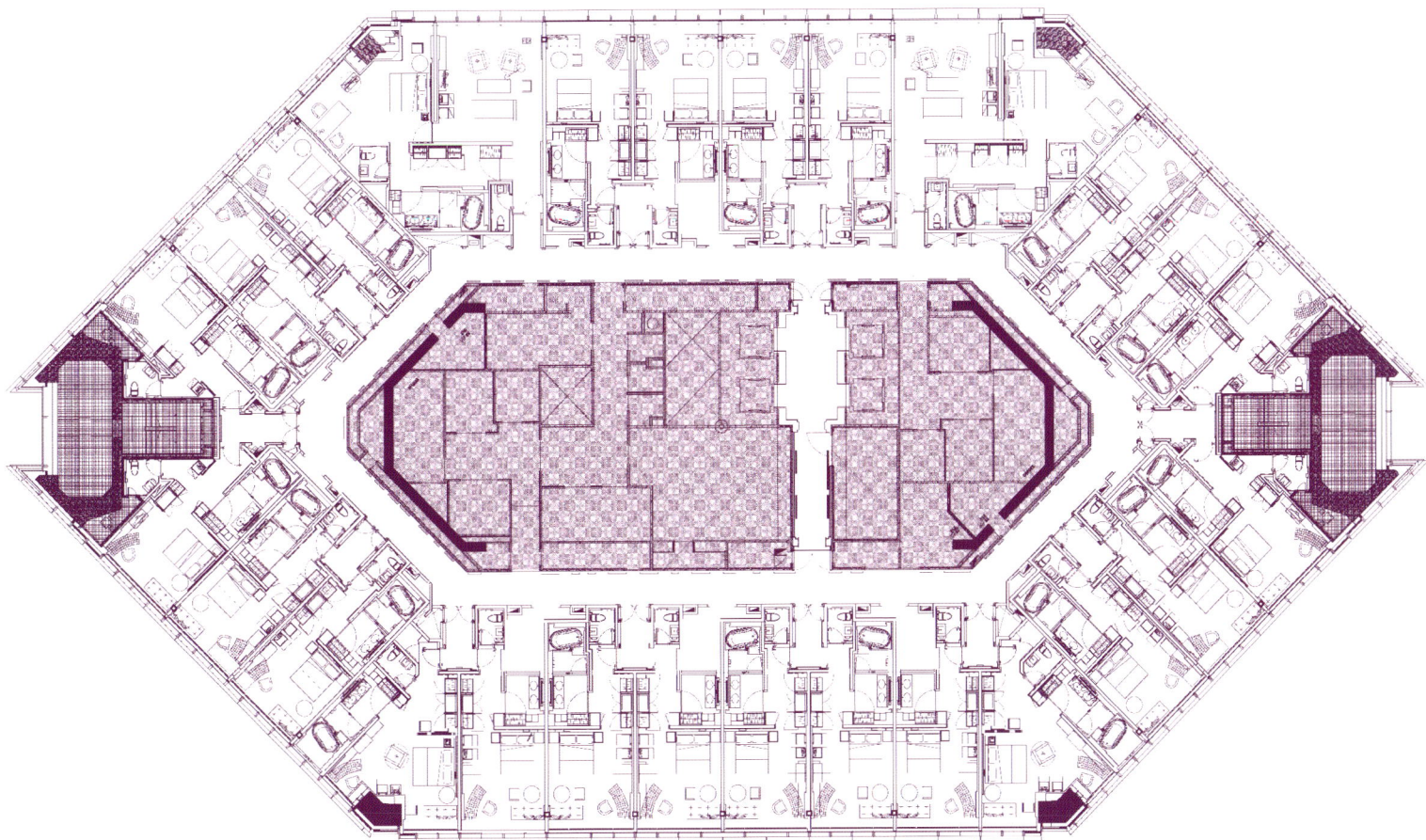

↑↓85层游泳池 / 85th floor swimming pool.
→游泳池楼梯 / Swimming pool stairs.

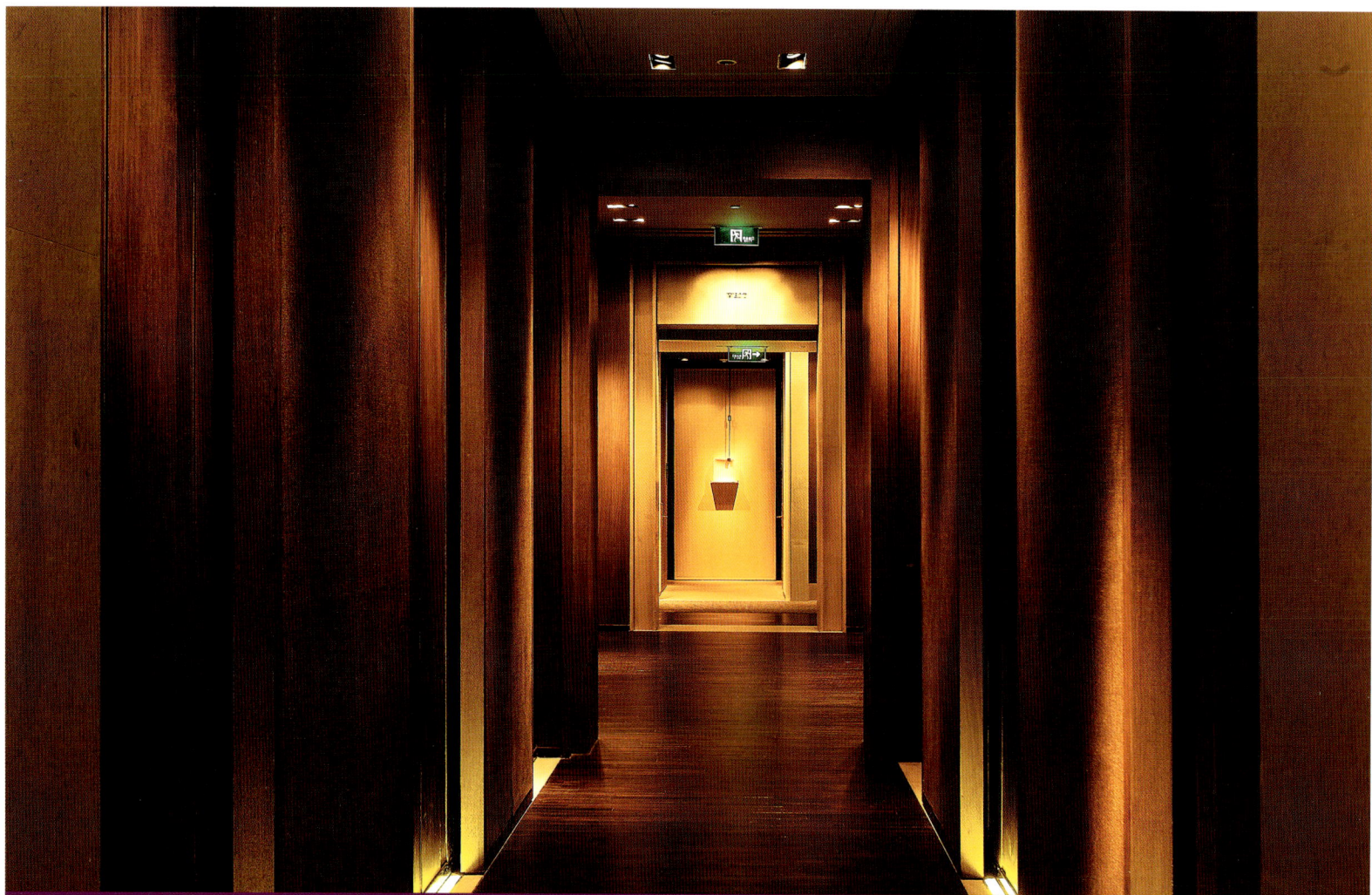

↑86层会议室外的通道 / 86th floor of the corridor outside the meeting room.
→86层会议室 / 86th floor conference room.
↓86层私人宴会厅 / 86th floor, a private banquet rooms.

↑91层的餐厅入口 / 91st floor restaurant entrance.
←餐厅内的吧台 / The bar inside the restaurant.
↓91层的餐厅 / 91st floor restaurant.

← ↑ 92层的餐厅 / 92nd floor restaurant.
↓ ↓ 餐厅上方的装饰品 / Restaurant at the top of the ornament.

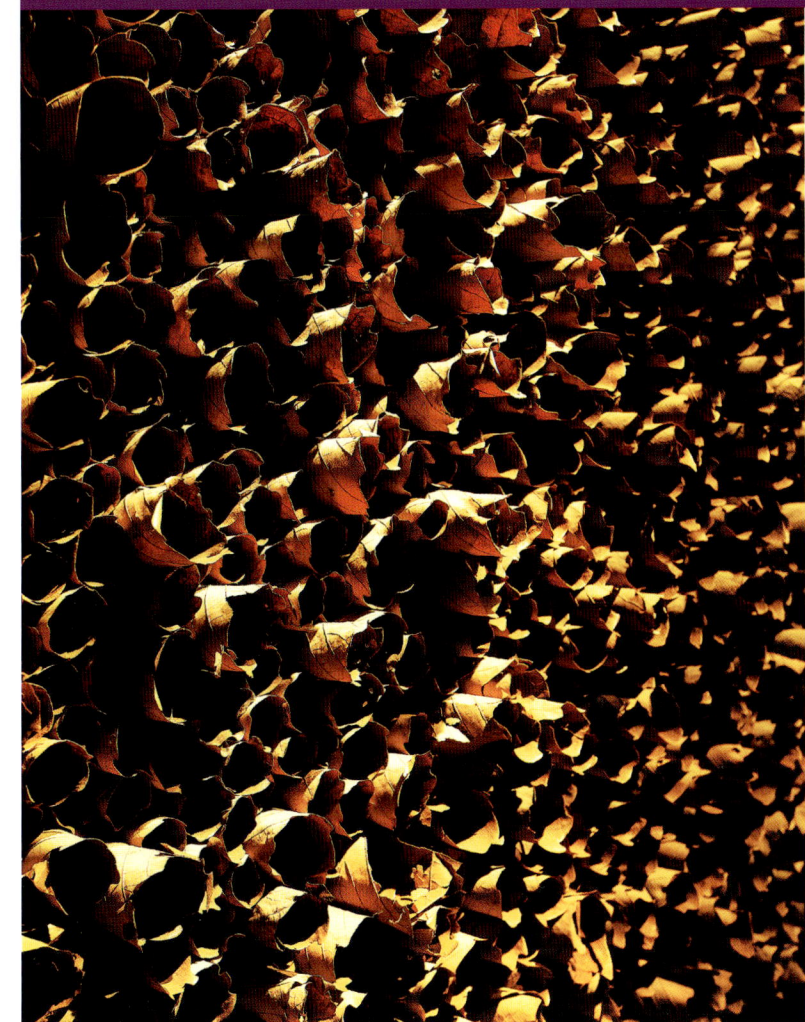

↑92层的中式酒吧入口 / 92nd floor Chinese-style bar entrance.
→↓92层的中式酒吧 / 92nd floor Chinese-style bar.

↑92层西式酒吧 / 92nd floor Western-style bar.
→93层西式酒吧 / 93rd floor lounge decorated deer.

CARBON
HOTEL

Carbon酒店

07

【工程名称】Carbon酒店
【坐落地点】比利时根克
【面积】7000 m²
【设计】Peter Cornoedus and Partners 〔比〕

整个酒店共有60间客房，房间面积大小不等，一条宽敞的走廊将所有的客房与街道隔开，降低了外部噪音进入房间内的可能性。室内的墙面采用高品质油漆和艺术墙纸两种材料，不同的光泽和质地在墙面上产生奇妙的对比。

所有的客房中，由于厕所的私密性较强被围合起来，除此之外其余的部分都被安排在同一个空间里。浴室区域通过一个悬挂在天花上的半截罩子和升起的平台来确定。在面积大一点的客房中，浴缸直接嵌入白色的床头板中，与床连成一体。在小一些的客房中，浴缸和床铺上方的下落式天花增进了空间的私密感。

Carbon酒店的餐厅是整个设计中的亮点。尤其是餐厅中悬空吊顶是空间中最具特色的部分，来自摩洛哥的水泥砖被绘上各种不同图案，拼贴出来的样子十分花哨，最有趣的是这些图案相互之间没有任何关联，无论色彩，还是原产地，好像与酒店都格格不入，但是营造出来的奇特效果却非常与众不同。

酒店顶层的水疗中心将黑白两种色彩演绎到极致，总面积超过700 m²，拥有6个治疗室，其中还体贴入微地设计了几处娱乐消遣区域。

A total of 60 rooms throughout the hotel, room size ranging from a spacious corridor will be separated from all of the rooms and streets, reducing the external noise into the room possibilities. High-quality interior wall paint and artistic wallpaper two kinds of materials, different gloss and texture of the wall, wonderful contrast.

All the rooms, due to strong privacy toilet besieged together, apart from the rest of the parts are arranged in the same space. Bathroom region through a hanging ceiling of the half of it urges hood and the rising platform to determine. In an area bigger room, the bath directly embedded in the white headboards, and bed fused. In the smaller room, bath and beds at the top of the whereabouts of type ceiling enhance the privacy of a sense of space.

Carbon hotel's restaurant is a bright spot in the entire design. In particular the vacant restaurant space in the ceiling is the most unique part of the cement blocks from Morocco has been decorated with a variety of different patterns, collage out like a very fancy, the most interesting is that these patterns there is no correlation between them, no matter the color, or country of origin, it seems incompatible with the hotels, but managed to create a bizarre effect that is very different.

The top floor spa hotel will be two kinds of black and white color interpretation to an extreme, a total area of more than 700 m², has six treatment rooms, which also designed an enhanced level of several entertainment areas.

↑马赛克贴面的接待台 / Mosaic veneer reception station.

086

↑餐厅入口处"碳"的主题 / Restaurant at the entrance to "carbon" theme；↑酒店公共区域 / Hotel public areas.
→酒店餐厅，宽条防腐木地板铺地 / The hotel's restaurant, wide wooden floor paving of corrosion.
↓餐厅装饰细部 / Restaurant decorative detail；↓餐厅俯视角度 / Restaurant overlooking the point of view.

↑客房 / Rooms

→客房，黑白两色占据空间主导地位 / Rooms, black and white two colors occupy the space for the leading position.

↓浴缸直接嵌入白色的床头板中 / Bathtub directly embedded in a white bed sheet.

↑ 垂下的天花罩住床铺和浴缸 / Hanging ceiling covering the beds and bathtub.
←抬起的地坪和垂下的天花划出浴室区域 / Raised ground floor and dropped ceiling to draw the bathroom area.
↓SPA室 / SPA room；↓SPA室局部 / Local SPA room.

091

CHRISTIAN'S
H O T E L

克丽司汀酒店

08

【坐落地点】河南洛阳；【面积】14 400 m²
【设计】王政强；【参与设计】刘牧、魏尚、王松峰、李君岩
【设计单位】郑州弘文建筑装饰设计有限公司
【摄影】贾方

酒店地处洛阳最繁华的商业中心"洛阳新都汇"，总面积为14400 m²，共15层，其中5至14层为客房楼层，客房数量为158间，其次是一个大堂、一个包含自助餐功能的餐厅，以满足不同客人的需求。

酒店净白的大堂将外面繁杂和喧闹隔开。大堂内以笑脸为主题雕塑般的柱子造型，时刻欢迎着入住酒店的客人。每层客房我们特定了不同的主题风格，为入住的客人增添趣味性和神秘感。有改变原有空间的中规中矩，以30°基角、长方体和圆形为变化组合的个性动感楼层；30°、立方体、圆趣；有安静、平和令人忘却尘世烦恼、放松愉悦；象牙塔、四月天、挪威森林；还有结合洛阳古都特点及现代人的需要提炼出梦回唐朝情感的"唐韵"楼层和既拥有人们向往充满活力新古典主义风格的"罗马假日"楼层；"透明时代"这一层用彩色玻璃规划出一个个看似透明实则私密的立体空间，透明简单却五光十色；而"云中漫步"这一层处在楼层最高端，以自然朴素的木质肌理结合清雅的色彩营造出一方东南亚的云上天空，这是一个让人沉思的空间，有如一方净土可以肆无忌惮的放松和休息，也可以随心所欲的冥想，自由体验飘浮于云端的感觉。

Luoyang hotel is located in the most prosperous commercial center "Luoyang new capital sink", with a total area of 14400 m², a total of 15 layers, of which 5 to 14 floor room floor, room number of 158, followed by a lobby, a buffet features include The restaurant, to meet different customers needs.

Whitening of the hall will be outside the hotel complex and noisy separating the lobby to smile for the styling theme of sculpture-like columns, always welcome to stay at the hotel guests. Rooms on each floor a different theme of our particular style, for the guests to add interest and mystery. A change in the original space, law-abiding, to 30 ° base angle, rectangular and circular dynamic personality combined for change floors: 30 °, the cube, round interest; a quiet, peaceful people on earth to forget troubles and relax pleasure: the ivory tower April day, the Norwegian forest; as well as combined the ancient capital of Luoyang, characteristics and needs of modern people to extract the emotional Tang Dynasty "Tang Yun," floors, and both have people yearning for a vibrant neo-classical style "Roman Holiday" floors; "Transparent Time "This is a layer of colored glass with a seemingly transparent work out a private but in reality three-dimensional space, simple but colorful and transparent; and" Walk in the Clouds "This is a layer at the floor of the most high-end, natural simple wood texture with elegant color to create a cloud on the sky one of Southeast Asia, it is a contemplative space for people, like the Pure Land one can do anything to relax and rest, meditation can also be arbitrary, free to experience the feeling of floating on clouds.

↑酒店的建筑外观 / Architecture of the hotel exterior.

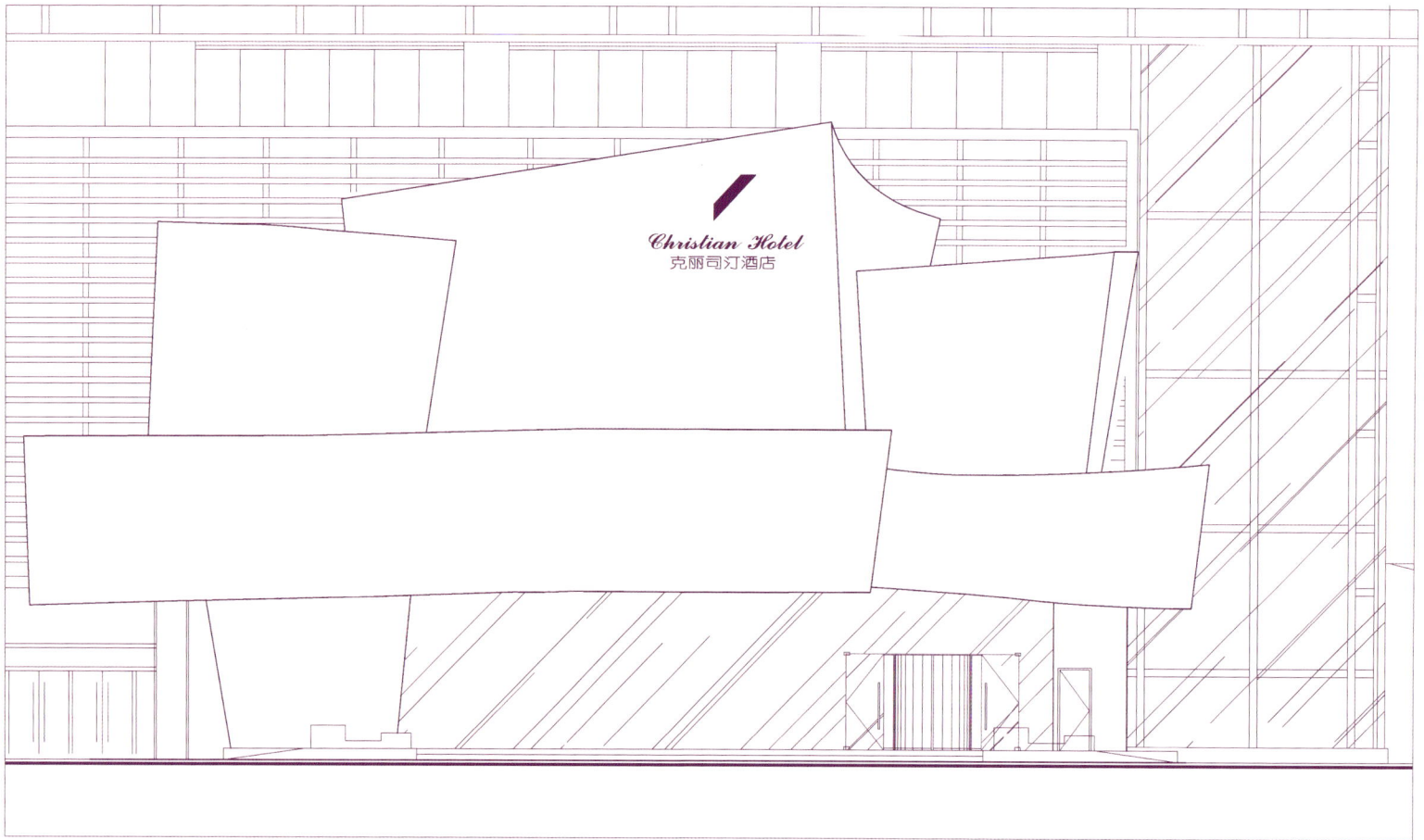

Christian Hotel
克丽司汀酒店

↑入口外立面 / Entrance facade.
↓酒店净白的大堂 / Whitening in the lobby of hotel.

↑↓人坐内以笑脸为主题雕塑般的柱于造型，时刻欢迎着入仕的各人 /

Lobby to smile for the styling theme of sculpture-like columns, always to welcome guests.

095

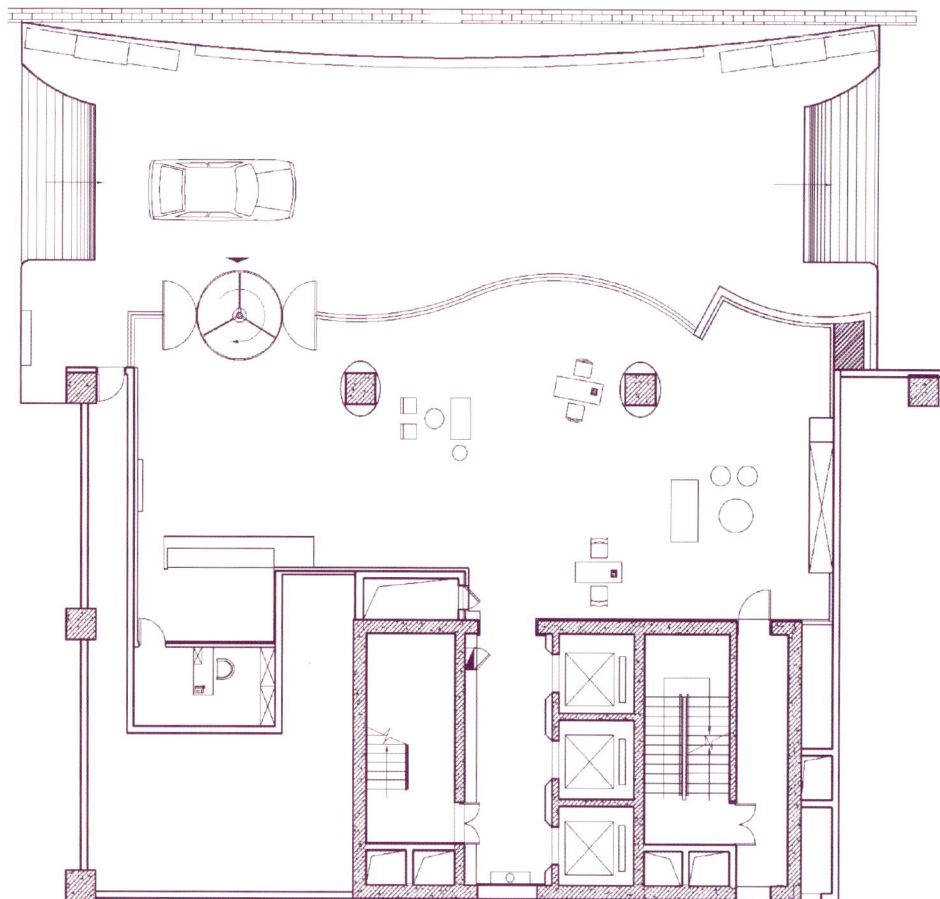

↑大堂平面图 / Hall floor plan.
→↓ "圆趣" 走廊 / "Yuan Qu" corridor.

↑ "圆趣" 客房内部 / "Round Fun" room inside.
↓ 客房内部的布局和陈设 / Room layout and furnishings inside.

↓ "圆趣" 平面图 / "Yuan Qu" plan.

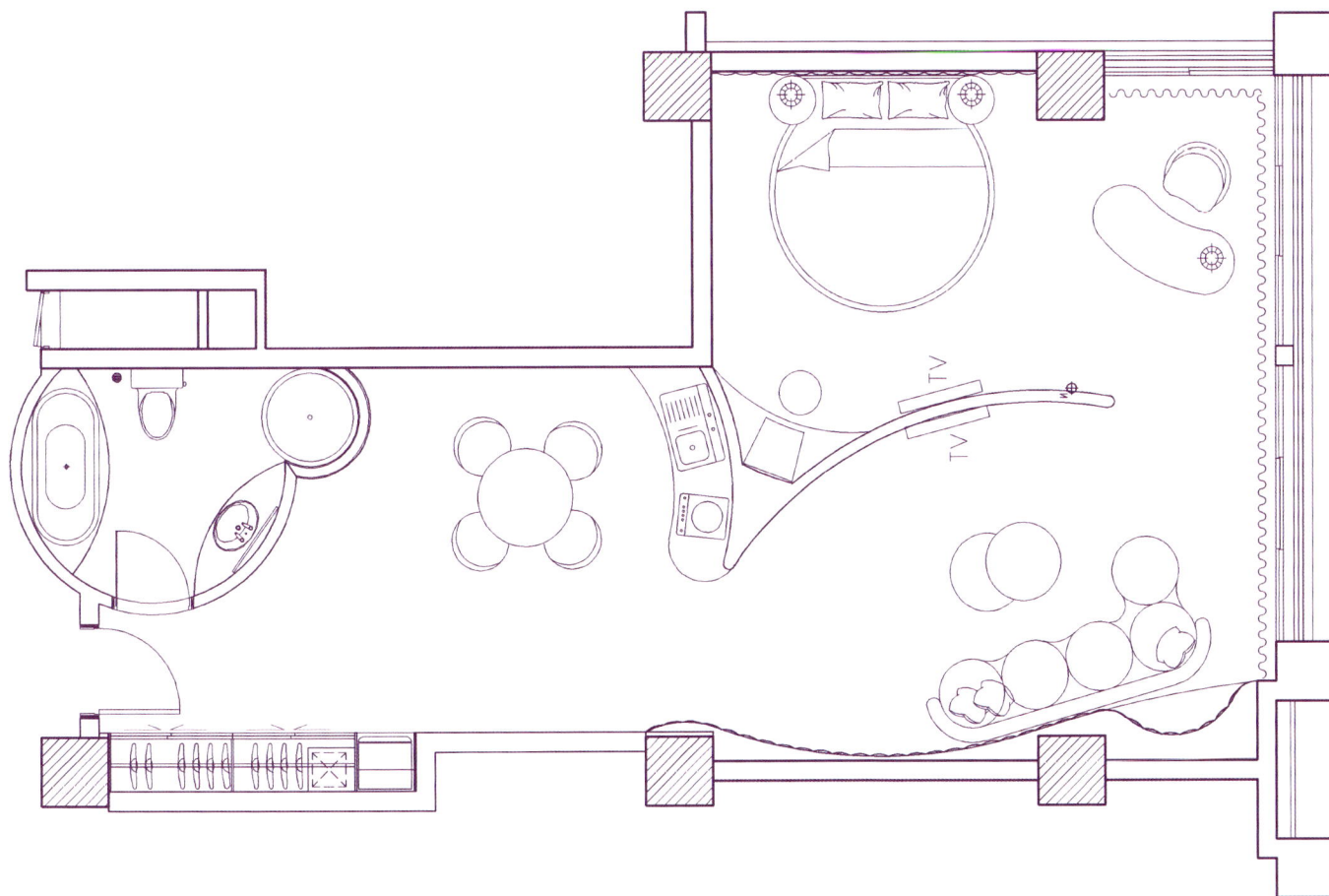

↑ "挪威森林"客房内部 / "Norwegian Forest" room inside.
↓ 客房的起居室 / Rooms of the living room.

↓ "挪威森林" 平面图 / "Norwegian Forest" plan.

餐厅

客厅

卧室

卫生间

主卫

↑ "云中漫步" 客房内部 / "Walk in the Clouds" room inside.
↓ "云中漫步" 平面图 / "Walk in the Clouds" plan.

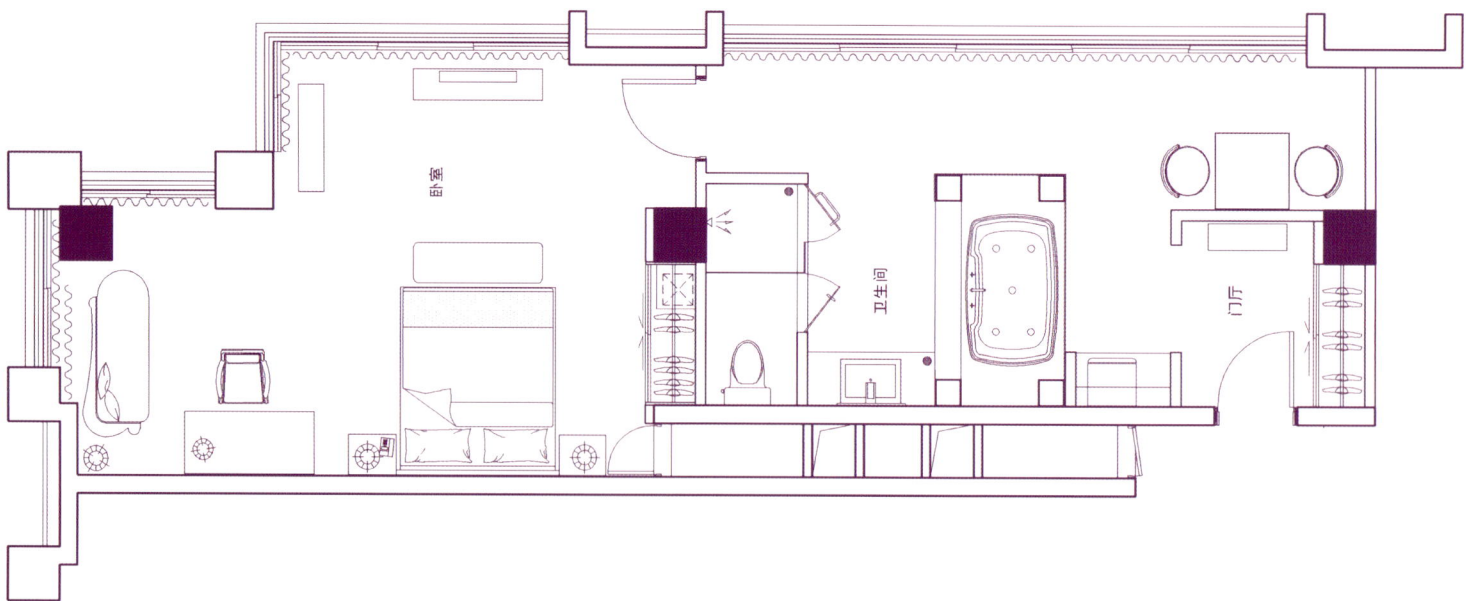

卧室

卫生间

门厅

↑ "罗马假日"客房内部 / "Roman Holiday" room inside.
↓ "罗马假日"平面图 / "Roman Holiday" plan.

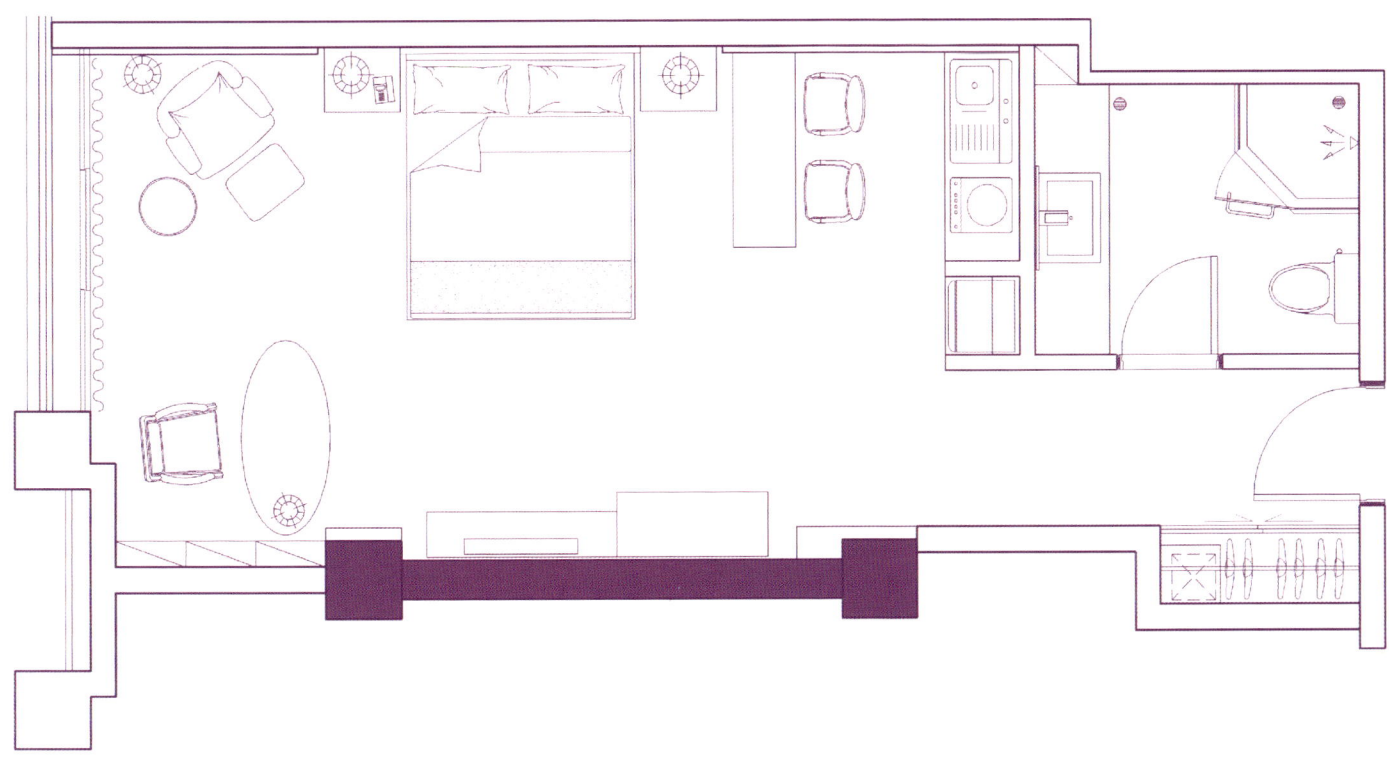

CITIZEN M
HOTEL

CitizenM酒店

09

【坐落地点】Jan Plezierweg 2，118 BB，Schiphol机场，荷兰；【面积】6000 m²
【设计】Concrete建筑设计事务所〔荷〕
【参与设计】Rob Wagemans、Erikjan Vermeulen、Jeroen Vester、Erik van Dillen 、Matthijs Hombergen
【摄影】Richard Powers

本案是世界上第一家CitizenM旅店，穿过一个红色的透明盒子就进入了旅店的大堂。实际上，将这个空间称之为"客厅"更为恰当，因为这里没有奢华酒店的富丽堂皇，迎面而来的是无比的温馨舒适。

一层的公共区域被划分成若干个不同的部分，包括工作区、餐厅和休闲区。旅店共有230间客房，每间客房都只有14 m²，房间内的设计毫无二致，让人无法不承认，这就是现代生活的完美体现。

客人在办理入住手续时，便可以通过自助服务终端来对客房进行个性化设计，包括音乐、灯光、温度等，也可以在入住客房后在房间中的触摸屏键盘来设置，同时还可以享受到免费的互联网接入和点播式电视服务。

为了充分利用空间，卫生间采用了分离设计，两个巨大的玻璃圆筒各占一边，一个是淋浴间，一个是厕所。还有一个较小的圆柱体则容纳了洗手池和储藏空间，并与其他空间分割开来。淋浴间采用全透明的玻璃，一半被固定，一半是两扇移门，客人可以根据自己的喜好使用两种不同的淋浴设备。淋浴区用幕布隔开，避免水汽扩散；厕所相对来说更具私密性，因而使用的是磨砂玻璃，功率强大的通风系统为这个小空间解除了如厕时带来的尴尬。

Case was the first one CitizenM hotels, through a transparent red boxes entered the hotel lobby. In fact, this space will be called the "living room" is more appropriate because there is a magnificent luxury hotel, greeted by the warmth unparalleled comfort.

The ground floor public area is divided into several different parts, including the work area, restaurants and leisure areas. A total of 230 hotel guest rooms, each room had only 14 m², the design of the room identical, people can not recognize, this is the perfect embodiment of modern life.

When guests check-in procedures can be through self-service terminals to the room to personalize the design, including music, lighting, temperature, etc., you can also stay in the room after room of the touch-screen keyboard to set up, but also be able to enjoy free of charge Internet access and on demand TV services.

In order to take advantage of space, separate bathroom with a design of two huge glass cylinder each side, one shower, one toilet. There is a smaller cylinder to accommodate the hand-washing pool and storage space, and with other space-separated. Shower all-transparent glass, half fixed, half of the two-shift gate, guests can according to their own preferences using two different showers. Shower areas are separated with curtains to prevent water vapor diffusion; toilets are relatively more privacy, and therefore using a frosted glass, powerful ventilation system for this small space to lift the toilet when the embarrassment caused.

↑酒店的建筑外观 / Architecture of the hotel exterior.

↑整个建筑看起来像一个大大的金属盒子 / The building looks like a big metal box.
↓外墙剖面图 / Wall section.

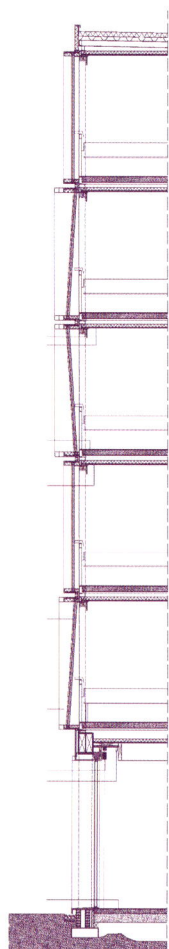

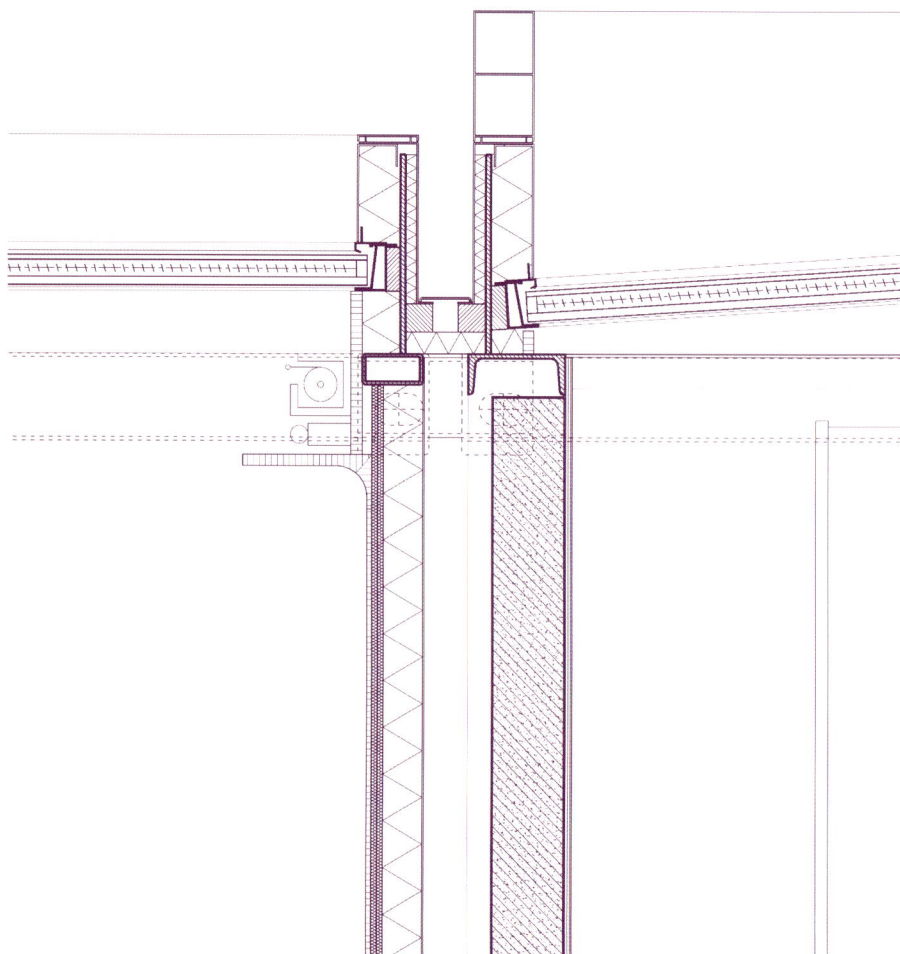

↑建筑立面图 / Building main view.
↓一层平面图 / 1st floor plan.

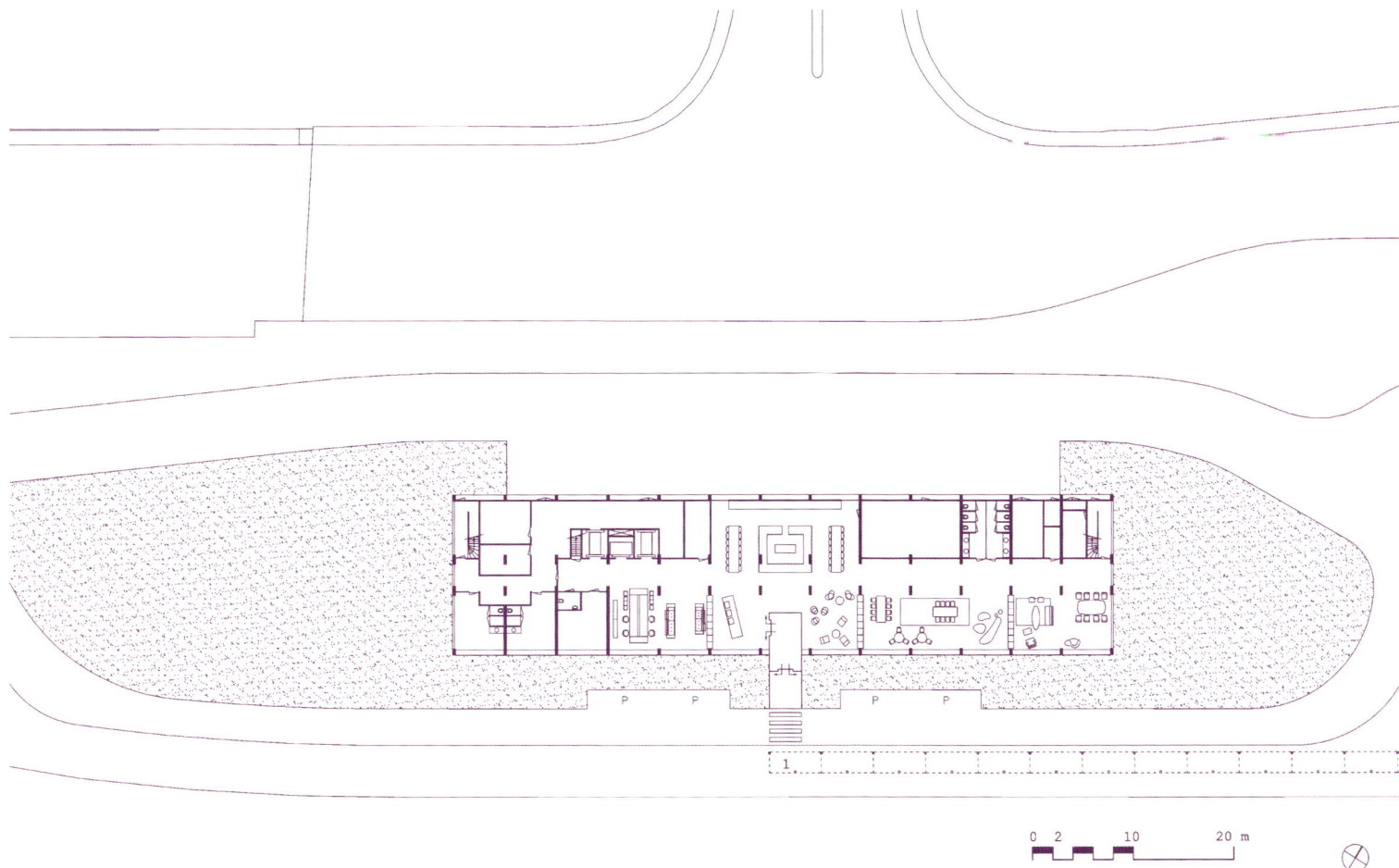

0 2   10   20 m

↑↓↓餐厅的中央是自助服务区，客人可以挑选喜欢的食物和小商品 /
The central is a self-service restaurant area, guests can choose from a selection of food and commodities.

↑休闲区的装饰富有设计感 / Leisure areas, the design sense of the rich decoration.
↓标准层平面图 / Typical floor plan.

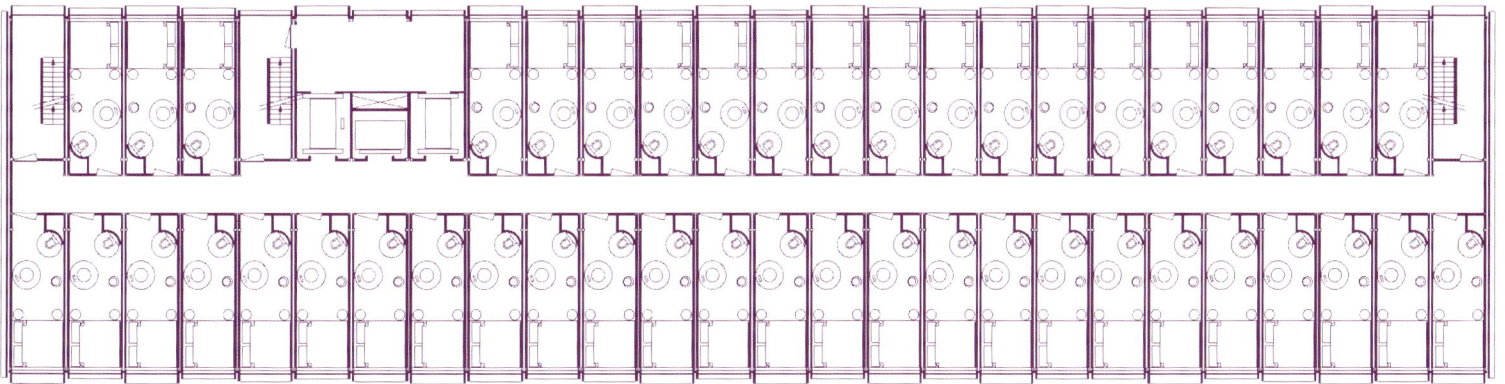

↑ 客房内部设计 / Room interior design.
← 客房内，大大的窗户采光充分 / Room within the large windows increase natural lighting.
↓ 两个巨大的玻璃圆筒分别是淋浴间和厕所 / Two large glass cylinder are showers and toilets.

↑为避免干扰，会议厅是围合起来的 / In order to avoid interference, the conference room was enclosure up.

113

W HOTEL
IN HONG KONG

香港 W hotel

10

【坐落地点】香港九龙
【设计】深田恭通 〔日〕、Nie Graham 〔澳〕
【设计公司】GLAMOROUS 设计公司、g+a设计公司
【摄影】(c)Nacasa & Partners

香港的W Hotel 贯穿W品牌独有的〝随时、随需〞的服务精神，提供贴身服务。酒店正面玻璃幕墙内侧，无数根金粉包裹的小树枝组成的抽象森林，悬挂在玻璃幕墙内侧。配合着LED灯的闪烁，巨大的〝W〞标志向香港——这个远离森林的城市，展示着自己的魔力。

酒店的一层是迎宾区，室内的装饰传达着〝大地能量〞这一主题：墙面菱形的方格描述五色泥土让岁月凝固，酒店的标志被艺术家演绎成LCD显示屏播放的燃烧火焰来展现温暖和无穷的能量。

一层通往六层的大堂的观光电梯内，用超现实主义的手法来表达自然中水和木的关系。透过轿厢正面的玻璃，客人可以近距离的看到悬挂在空中的金色小树枝，和树枝后面香港的美景。电梯门开启，客人们又进入到另一个场景：Living Room。

六层的大堂被设计师装饰成为一个舒适的起居室；七层的宴会大厅的主题是〝大地的富饶〞；商务中心和会议室在八层，用横向交错的软织墙面和绿色地毯来表达〝抽象的花园〞这一主题。客房层电梯间和走廊是客人从公共区过渡到安静私密的客房的地方。

72层水疗中心bliss的室内装饰围绕着〝绿水〞的概念展开。水是生命和能量的源泉，也是SPA的精神所在。

W Hotel in Hong Kong throughout the W brand's unique "anytime, on-demand" service spirit, providing personal services. Hotels inside the front glass curtain wall, and numerous small branches root powder wrapped in an abstract composition of the forest, hanging inside the glass curtain wall. Combined with the LED lights flashing, huge "W" logo to the Hong Kong - a city far from the forest, and demonstrates the magic of their own.

Hotel layer is a welcome area, interior decoration to convey the "earth energy" this theme: diamond-shaped wall of the box description of the soil for years colored solidification, the hotel is interpreted as a sign of being an artist LCD screen display to the burning flame show warmth and endless energy.

Layer leading to the six-story elevator inside the lobby of the tourist, using ultra-realistic way to express the nature of the relationship between water and wood. Through the glass front elevator, guests can be close to see that hanging in the air, the golden beauty of the small branches and back. Elevator doors opened, guests entered into another scene: Living Room.

Six-story lobby has been designer decorated as a comfortable living room; seven of the banquet hall's theme is "Earth's rich"; business center and conference rooms on the eighth floor, with staggered horizontal soft woven walls and green carpet to expression "abstract garden" this topic. Room floor elevator and hallways are like customers transition from the public area of the room quiet and private place.

72-storey spa's interior around the bliss of "green water" concept started. Water is the source of life and energy, but also the spirit of the SPA.

↑酒店的建筑外观 / Architecture of the hotel exterior.

↑酒店正门的巨大的"W"标志 / Hotel main entrance of the big "W" logo.

↑一层迎接大厅传达着"大地的能量"这一主题 / A layer of communication to meet the hall of "earth energy" this topic.

↑设计师在此创造了一种低调奢华的美感 / This designer has created a low-key luxury and beauty.
←墙面菱形的方格展现温暖和无穷的能量 / Wall diamond-shaped grid to show warmth and endless energy.
↓每一个细节都透露出W hotel的精致不凡 / Revealed every detail of the exquisite W hotel extraordinary.

↑七层的宴会大厅的主题是"大地的富饶" / Seventh floor of the banquet hall's theme is "fertile earth."
↓餐厅的包间 / Dining rooms.

↑西餐厅的餐桌 / Western restaurant tables.
↓开放西餐桌 / Open Western dinner table.

↑客房里到处洋溢着自然的活力 / Everywhere room filled with natural vitality.
↓暖暖的客房 / Warm rooms.

↑客房的故事 "大自然的火花" / Rooms of the story "sparks of nature."
↓客房的卫浴 / Room bathroom.

↑八层是商务中心和会议室 / Eighth layer is the business center and conference rooms.
↓电梯间里电梯的按钮巧妙地隐藏在书架中 / Lift buttons inside the elevator hidden in the shelves inside.

↑会议室 / Conference room.
↓72层水疗中心 / 72nd floor spa.

the 'youth' will set you free
you look 'spa-velous', darling
all systems 'glow'
create your best 'face' scenario
hold the 'lines', please
it's the 'taut' that counts
take the 'spa' into your own hands
a 'soak' of genius
desperately seeking 'smoothin'?

↑水疗中心的楼梯 / Spa stairs.
↓↓水疗中心bliss的室内装饰围绕着"水绿色"的概念展开 / Spa bliss of the interior decoration around the "water green" concept started.

↑ 舒适的环境让客人在这里得到最好的休息 / Comfortable environment for the guests get the best rest.
↓ 水是生命和能量的源泉 / Water is the source of life and energy.

SANCTUM
SOHO HOTEL
Sanctum Soho酒店

11

【坐落地点】英国伦敦
【工程名称】Sanctum Soho Hotel
【设计】Lesley Purcell　Can Do 〔英〕
【建筑设计】Smith Caradoc-Hodgkins

这家酒店由两个格鲁吉亚排屋组成，大堂的装饰画、老式灯具，两个壁炉，甚至连房门的锁都返朴到用钥匙而非现代化的电子门卡，一派老式英伦风情。传统总给人以亲切感，以建筑来表达传统，展现一地文化，为城市增加应有的识别度是设计的重要意义。

客房的30个房间各有千秋，从最小的房间、套房到顶层公寓，没有一间是重复的。设计师精心挑选了4种颜色作为设计主题，以迎合不同客人的口味：有阳刚气十足的深朱古力色、流光溢彩的银色、感性的粉红色和富有异域风情的紫色。

进入房间后，古朴之下的现代化让世人乐享客居酒店的生活。先进的娱乐系统、iPod的连接器、Wii游戏机、免费Wi-Fi接入，满足每一个新潮青年的需求。同时，房间内提供伦敦著名品牌REN Skincare的护肤品，床上用品是高级埃及棉，毛巾、浴袍及拖鞋使用意大利奢侈品品牌Frette的产品，都为客人提供一个最佳的环境。酒店内还设有健身房和屋顶水疗中心，餐厅提供经典英式菜肴，酒吧向会员和客人最大限度地提供高品质的食物和酒水。Sanctum Soho Hotel还拥有私人电影院，十足的文艺风范，还可以根据不同的需求变化成活动场地。

The hotel consists of two Georgian townhouses composition, lobby decoration painting, old-fashioned lamps, two fireplaces, and even door locks have been traveling to the Park, instead of using modern electronic door key card with one of the old England style. Total gives a traditional affinity to the construction to express the traditions and culture to show a way for the city should increase the degree of recognition is to design importance.

Rooms 30 rooms is different, from the smallest rooms, suites to the top floor apartment, no one has been repeated. Designers have been carefully selected four kinds of color as a design theme, to suit the different tastes of guests: There are masculine aggressiveness deep chocolate color, bright colors and silver, sensual exotic pink and rich purple.

Entered the room, is to allow hotel guests to enjoy live music. An advanced entertainment system, iPod connector, Wii games, free Wi-Fi access to meet the needs of each new wave of young people. At the same time, the room to provide London's famous brand REN Skincare skincare products, high-ranking Egyptian cotton bedding, towels, bathrobe and slippers to use the Italian luxury brand Frette products are available for the guests the best environment. The hotel also has gym and rooftop spa, restaurant offers classic British cuisine, bar members and guests to maximize the delivery of high-quality food and drinks. Sanctum Soho Hotel also has a private cinema, full of literary and artistic style, you can also change according to different requirements into the venue.

↑ 阳刚气十足的深朱古力色系房间 / Masculine room full of deep chocolate color.
↓ 充满异域风情的紫色系房间 / The department of exotic purple room.

↑感性的粉红色系房间 / Perceptual color pink room.
↓流光溢彩的银色系房间 / Ambilight silver line the room

↑黑白色的运用使得空间布局分明 / Black and white makes the use of distinct spatial distribution.
↓深色墙面搭配花纹小椅，减弱空间的厚重感 / Dark-colored walls with patterns of small chairs, less heavy feeling of space.

↑ 半圆座位增加空间的私密性 / Semicircle seating with additional space for privacy.
↓ 餐厅一角 / Restaurant corner.

HYATT
ON THE BUND
上海外滩茂悦大酒店

12

【坐落地点】上海北外滩黄浦江西面堤岸；【面积】100 000 m²
【建筑设计】美国H.O.K设计公司
【室内设计】Remedios Siembieda Inc
【施工单位】苏州金螳螂建筑装饰公司

酒店的风格在装饰设计上展示，表达的是一种联想和向往，整体风格简约而清新。主色调通过深黑檀木色的木饰面、精致的亚光银色饰件、亚麻黄的花岗岩糅合，从而使整个空间透出豪华而又不失沉稳的大气。共享大厅的弧形玻璃钢结构顶棚，增强室内现代感同时，将自然光源引入室内，丰富采光层次。

大宴会厅里32盏高低大小不一的叶形灯，造型独特而新颖。酒店的中餐厅与大堂相联，风格秉承。灯光处理是一门技术，更是艺术的体现。

酒店西楼顶部4层是非常时髦VUE餐厅。该餐厅由著名日本室内设计公司Super Potato担纲设计，其创意灵感源于单身贵族的多层家居概念。作为主用餐区的餐厅部分位于最下一层，整个区域连续地将温馨舒适的"客厅"、多姿多彩的"书房"、宽敞明亮的"厨房"融为一体，处处营造居家的惬意氛围。在2270 m²的空间里，私人住所的概念流畅地在空间中体现；窗外的视野里，融合了外滩拥有历史感的建筑和浦东现代感十足的高楼大厦。

Style of the hotel design in the decorative display is a kind of association and expression of yearning, the overall style is simple and fresh. Primary colors through the dark ebony-colored wood finishes, exquisite matt silver ornaments, linen blend of yellow granite, so that the entire space revealing a luxury but without losing the atmosphere calm. Sharing the hall ceiling curved glass, steel, enhance indoor modern at the same time, the introduction of the indoor natural light, rich lighting levels.

Grand Banquet Hall 32 the level of large and small leaf light, unique and novel form. The hotel's Chinese restaurant and the lobby linked style uphold. Light treatment is a technology, and more art is all about.

Hotel at the top of four layers is a very stylish restaurant VUE. The restaurant by the renowned Japanese design firm Super Potato featuring interior design, its creative inspiration from the concept of singles in multi-storey home. As part of the main dining area of the restaurant is located in the most the next level, the entire region will be continuously warm and comfortable "living room", the colorful "study", bright and spacious, "Kitchen" integrated, everywhere to create a cozy atmosphere of home . In the 2270 m² of space, the concept of a private residence smoothly reflected in the space; the window of vision, the integration of the Bund has a sense of history full of buildings and modern skyscrapers of Pudong.

↑酒店共享空间，将自然光源与水景引入室内 / Hotel sharing space, natural light source and the introduction of the indoor water features.

135

↑ 旋转楼梯 / Rotate the stairs.
↓ 自然的空间 / The natural space.

↑一层大厅 / Floor lobby.
↓一层平面图 / Floor plan.

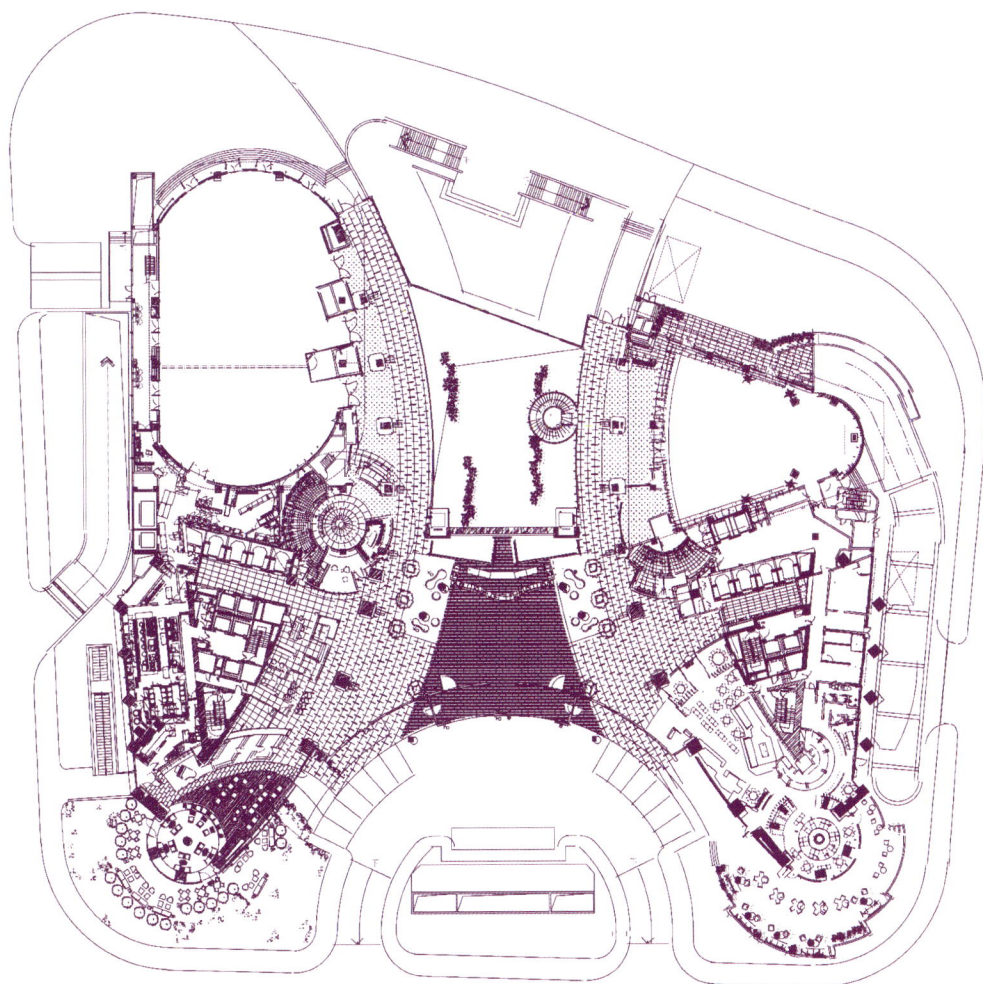

138

↑酒店的中餐厅 / The hotel's Chinese restaurant.
↓家一般的餐厅，勾起人温暖的记忆 / Restaurant at home, brought back warm memories of people.

↑酒店的宴会厅 / The hotel dining hall.
↓大型的会议室 / A large conference room.

VQ VENTIQUATTRO
HOTEL

VQ Ventiquattro酒店

13

【坐落地点】迪拜Marina
【设计】Studio Matteo Nunziati 〔意〕
【建筑设计】Studio Matteo Nunziati   KEO international Consultant, GTI〔意〕
【摄影】Beppe Raso

整个大楼有20层，包括152间客房、2个健身中心、1个SPA馆、2个户外游泳池和多个餐厅、美容馆、酒吧、大厅、购物中心。酒店的设计灵感来源于古埃及的纪念碑建筑，严肃，宏伟又不失精巧。

酒店的大厅有6 m高，超过30 m长，17 m宽，全部以灰色的大理石铺地，这也是传统的意大利的设计手法。墙面用同色的大理石贴面，三种规格的大理石任意的铺贴排列，营造出一种很柔和的装饰感。

大厅中央是2个定制的、通高到顶的装饰柜，皮革表面和高光漆面的木饰面相间。大厅中央是两块巨大的手工编织的地毯。窗帘也采用全落地的形式，在面料的选择上也以有浮雕感的织物为主。

酒店共有三种类型的客房可供选择：50 m²的标准间，70 m²的单人间和110 m²的双人间。单人间和双人间都自带厨房，房间内用一个很有个性的餐柜把厨房和生活区区分开来。不对称的线条勾勒出敞开空间的边界。

所有的客房都是很现代化的，整个房间弥漫的都是一种惬意和舒适的感觉，适当的装饰恰到好处，每个空间里都有一堵装饰墙壁用大尺寸的瓷砖饰面。

The entire building has 20 floors, including 152 rooms, two fitness center, a SPA Center, two outdoor swimming pools and several restaurants, beauty hall, bar, hall, shopping mall. The hotel's design takes its inspiration from ancient Egyptian monument building, serious, ambitious, yet delicate.

The hotel lobby there is 6 m high of over 30 m long, 17 m wide, all with gray marble flooring, which is the design of the traditional Italian way. With the same color of the marble wall veneer, three kinds of specifications Laying marble arbitrary order, to create a sense of a very soft decoration.

Center of the hall is a two customized, high-pass reach the peak of the cabinet, leather surfaces and high-gloss finish and white wood finishes. Center of the hall are two large hand-woven carpets. Curtain also used the whole floor in the form of choice of fabrics also have a sense of relief the main fabric.

There are three types of hotel rooms: 50 m² standard rooms, 70 m² of single rooms and 110 m² of double rooms. Single and double rooms have own kitchen, the room buffet with a great personality to distinguish between the kitchen and living District. Asymmetric lines outline the boundaries of open space.

All the rooms are very modern, the whole room is filled with the feeling of a cozy and comfortable, appropriate decorations just right, every space has a wall decorated with large size ceramic tile wall finishes.

↑酒店的建筑外观 / Architecture of the hotel exterior.

↑接待处，墙面和接待台用不同的石材拼贴 / A reception area, walls and reception station in different stone collage.
←酒店大厅，6 m的高度使空间显得气势宏伟 / 6m-high lobby of the hotel to make room look grand.
↓艺术的墙面 / Wall art.

143

↑游泳池的灯带，清晰地勾画出游泳池的边界 / Swimming pool light with a clear picture of a swimming pool borders.
↓游泳旁边的柚木凉棚 / Swim next to the teak arbor.

↑客房走廊 / Room corridor.

↑会客区的家具都出自意大利顶尖的家具公司 / Reception area of the furniture from Italy's leading furniture company.
↓客房空间 / Room space.

147

↑↑↓↓不同角度的客房，每一处均被用心的设计 /
Different angles of the rooms were carefully everywhere design.

↑ 厨房的小吧台 / Kitchen mini-bar.
← 厨房，酒店客房内还是很少见的 / Hotel room in a rare kitchen design.
↓ 卫生间的墙和地面采用同色的大理石 / The bathroom wall and floor had the same color of marble;　↓ 双人间的卫生间 / Double room bathroom.

LIMES
HOTEL
L

Limes 酒店

14

【坐落地点】澳大利亚布里斯班；【面积】1100 m²（单层为250 m²）
【设计】Alexander Lotersztain 〔阿〕
【建筑设计】Kevin Hayes Architects
【室内外设计】Derlot Pty. Ltd.

酒店的外观就是一个放大版的酒店LOGO，这一个强烈的酒店标志将出现在这个旅行的各个角落，包括一支铅笔、一本便签、一个洗漱包。酒店只有21间客房，另外还有时髦的咖啡馆、商店、酒吧、餐厅以及一个完全开放的屋顶酒吧、电影院。在庭院的阳台上，安装了吊床，有现代家具和舒适的床，空间与质感并重。

在保证功能的前提下，打破五星级酒店的设计框架，忠实地从旅行者基本需要出发，只想让你感觉最舒服。中规中矩的客房写字桌并没有出现，取而代之的是适合在膝上所用的笔记本电脑，甚至连椅子都省略了，可以直接坐在床边伏案工作。倒是房间里的小厨房是特意增加的功能，随时随地可以犒劳一下自己，厨房的椅子也可以用作办公椅，一椅多用。酒店提供全方位的无线上网，这也是酒店新的基本装备之一。衣柜的门板采用全镜面的质地，可以直接当穿衣镜用。客房的设计讲求简洁，垃圾箱、电缆等尽量减少或隐藏起来，使房间具有更清晰的视觉效果。

在材料的选择上，主要考虑了材料的耐久性、可维护性和外观性能，在温暖的心理暗示和现代材料的冷峻感之间找平衡。

Appearance of the hotel is a larger version of the hotel LOGO, which is a strong hotel logos will appear in every corner of this trip, including a pencil, one notes, a wash bag. Hotel only 21 rooms, in addition to trendy cafes, shops, bars, restaurants, and a completely open roof bars, movie theaters. In the courtyard of the balcony, installed hammock, with modern furniture and comfortable beds, space and texture of equal importance.

In the assurance function under the premise of the design framework to break the five-star hotels and faithfully from the basic needs of travelers, just want to make you feel most comfortable. Law-abiding, room desk, and does not appear, replaced by a suitable laptop used laptops, and even chairs were omitted, and can sit in bed and his desk. Daoshi room, a small kitchen is specifically added features can be anywhere Treat yourself, the kitchen chairs can also be used as office chairs, chairs a multi-purpose. The hotel offers a full range of wireless Internet access, which is a new hotel, one of the basic equipment. The door wardrobe with full mirror of the texture can be directly used when the full-length mirror. Room design emphasizes simplicity, trash, cable, etc. to minimize or hide so that the room has a clearer visual effect.

Hotels in choice of materials, the main consideration of the durability, maintainability and appearance of performance; in warm and modern materials, psychological suggestion solemn sense to find a balance between.

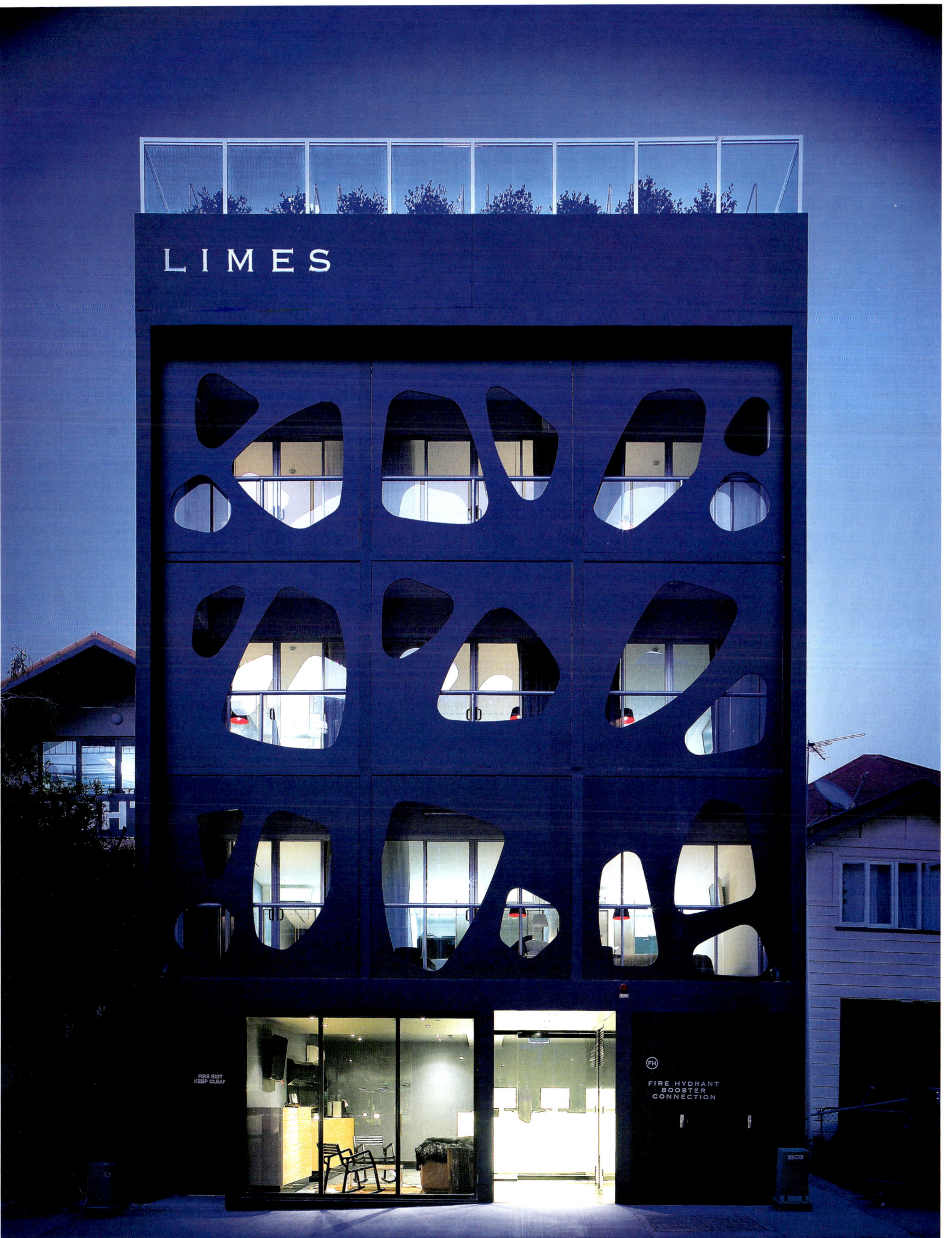

↑酒店的建筑外观 / Architecture of the hotel exterior.

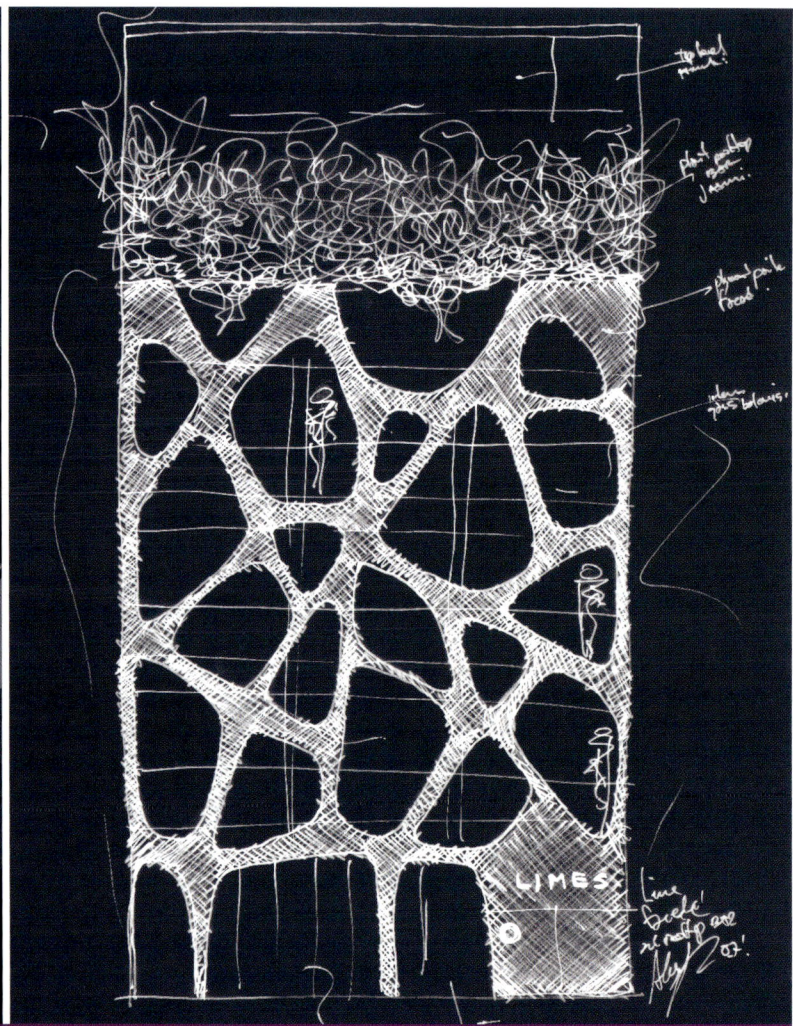

↑酒店的LOGO / Hotel LOGO; ↑Alexander手绘的酒店立面造型 / Alexander hand-painted facade of the hotel style.
↓标准间和套房平面图 / Standard rooms and suites plan.

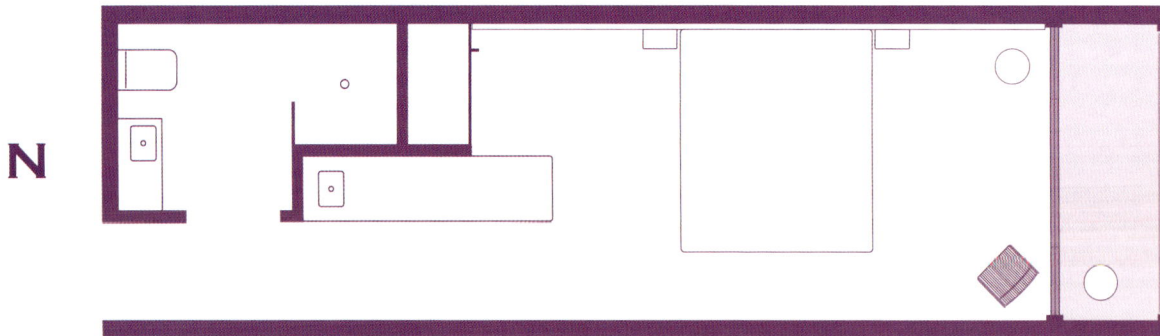

N

LIMES

1000
500

1:50

N

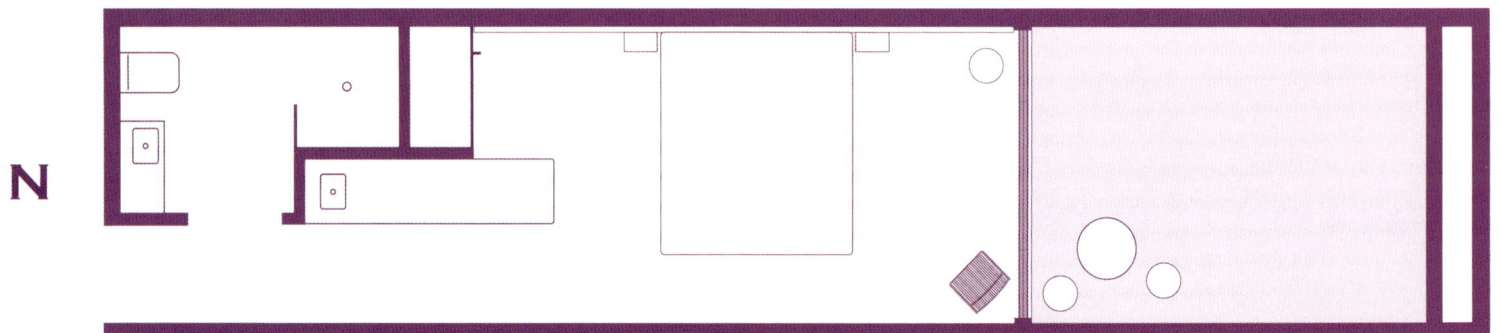

↑酒店入口处的休息区 / Hotel at the entrance to a rest area.
↓接待台 / Reception Desk.

↑床头小灯 / A small bedside lamp.
→客房内部，讲求有效面积的使用 / Room interior, emphasis on the use of the effective area.
↓餐椅与写字台 / Chair and desk；↓卫生间 / Bathroom

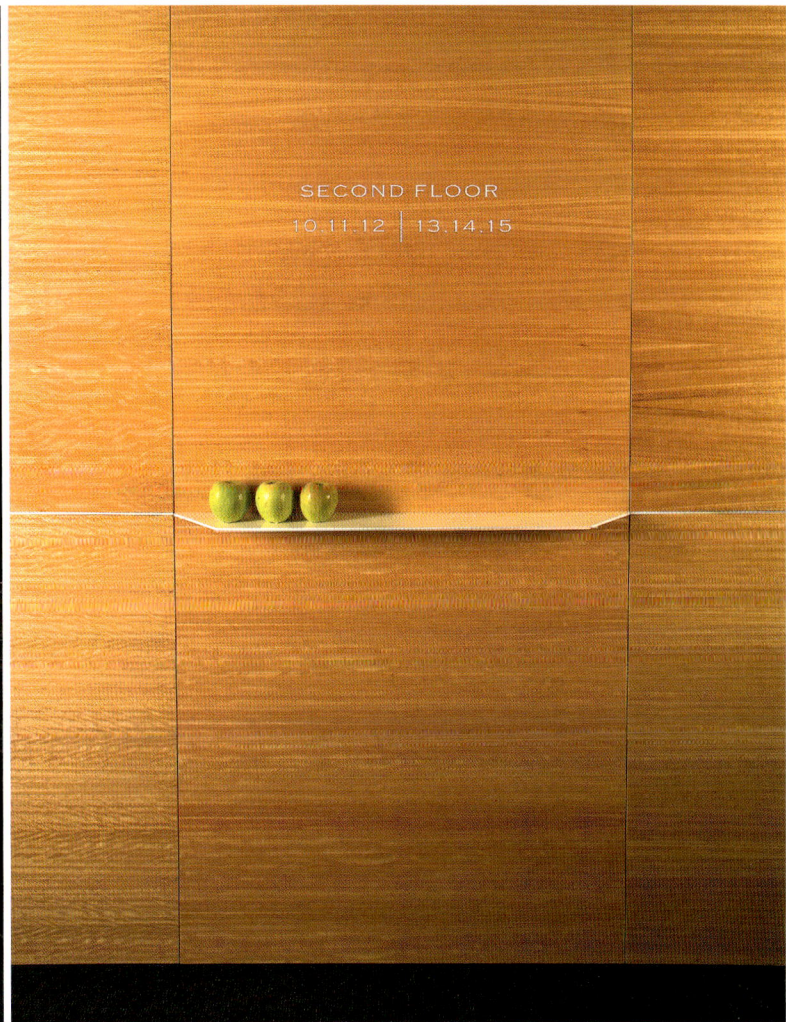

↑关照细节 / Attention to detail；↑走廊 / Corridor
←屋顶酒吧 / Roof bars.
↓冷灰里的一抹鲜艳，人见人爱的吊床 / The cold ashes of the touch of bright, cute and a hammock，↓庭院 / Courtyard

159

RESORT
SANYA BAY
三亚豪生度假酒店

15

【坐落地点】海南省三亚市三亚湾；【面积】110 000 m²
【设计】杨邦胜；【设计公司】杨邦胜酒店设计顾问公司
【主要材料】玛丽米黄、米黄洞石、海洋之星、贝壳马赛克、泰柚木饰面板、老船木雕刻板
【摄影】陈乙、文宗博

酒店室内建筑面积达11万m²，主楼客房1500间，度假别墅22栋，为目前中国最大五星级休闲度假酒店。除会议餐饮功能外，酒店整体设计吸收融合了东南亚建筑风格及热情的夏威夷风情，塑造出一种别样的热带风情，在椰风碧海处营造了一个真正关于"东方夏威夷"的纯美空间。

室内设计结合东南亚建筑风格及海南黎族文化，多文化的糅合拓宽了整体设计的内涵，展示三亚湾独特的风俗、文化和地域的魅力。空间大量采用"通透开敞"的手法来借景和展示，自然、质朴、通透、休闲，塑造出极具海南特色的度假酒店新形象，防腐防霉材料的大面积使用结合了当地的气候特点，更具美观和实用性。

在酒店大堂的整体设计中，室内设计保留了原有风貌，天花的设计沿用了建筑斜屋面结构，天花大面积采用密拼规格柚木板。柚木的重色降低了空间带来的空旷感，大堂上空整齐排列的木作格栅，横向与竖向延续了建筑斜屋面空间，既丰富了建筑的空间元素，又体现了海南度假酒店质朴、休闲的特色。整体设计在细节处的表达别有洞天，大堂内钢制的"鸟巢"装置，在一种灰黑色的整体色调中陈述低调的审美意识形态，打破了空间一以贯之的热带风情，注入了后现代的时尚元素。

Hotel interior construction area of 110,000 m², Main Room 1500, holiday villa 22, in order to present China's largest five-star leisure resort. In addition to meeting catering functions, the hotel combines the overall design of the absorption of Southeast Asia, Hawaii architectural style and warm atmosphere, create out of a different kind of tropical blue sea at the coconut to create a truly on the "Oriental Hawaii" the sweet space.

Interior design combines architectural style and Hainan Li Southeast Asian culture, multi-cultural blend of broadening the overall design of the content, display Sanya Bay's unique customs, cultural and geographical charm. Extensive use of space, "transparent open" approach, natural, simple, transparent, leisure, create a very unique resort of Hainan a new image, anti-corrosion anti-mold materials, the use of a combination of a large area of local climate characteristics, and more aesthetics and practicality.

In the hotel lobby's overall design, interior design retained the original appearance of the building followed the design of the ceiling inclined roof structure, ceiling to fight large-scale adoption of standard density teak panels, the heavy teak color reduces the open space to bring a sense of the lobby over the neatly arranged wooden grille, horizontal and vertical extension of the construction ramp roof space, not only enriched the architectural elements of space, but also reflects the pristine Hainan resort hotel and leisure features. The overall design is in the details at the expression of Journey Into Amazing Caves, steel lobby "Bird's Nest" device, in a dark gray color in the statement of the overall low-key aesthetic ideology, breaking the space-a consistent, tropical, injected elements of post-modern fashion .

↑酒店大堂 / Hotel lobby.

↑酒店景观，将室外的纯美风景引入室内 / Hotels landscape, the introduction of the indoor outside the sweet scenery.
→大堂接待台 / Lobby Reception Desk.
↓大堂平面图 / Hall floor plan.

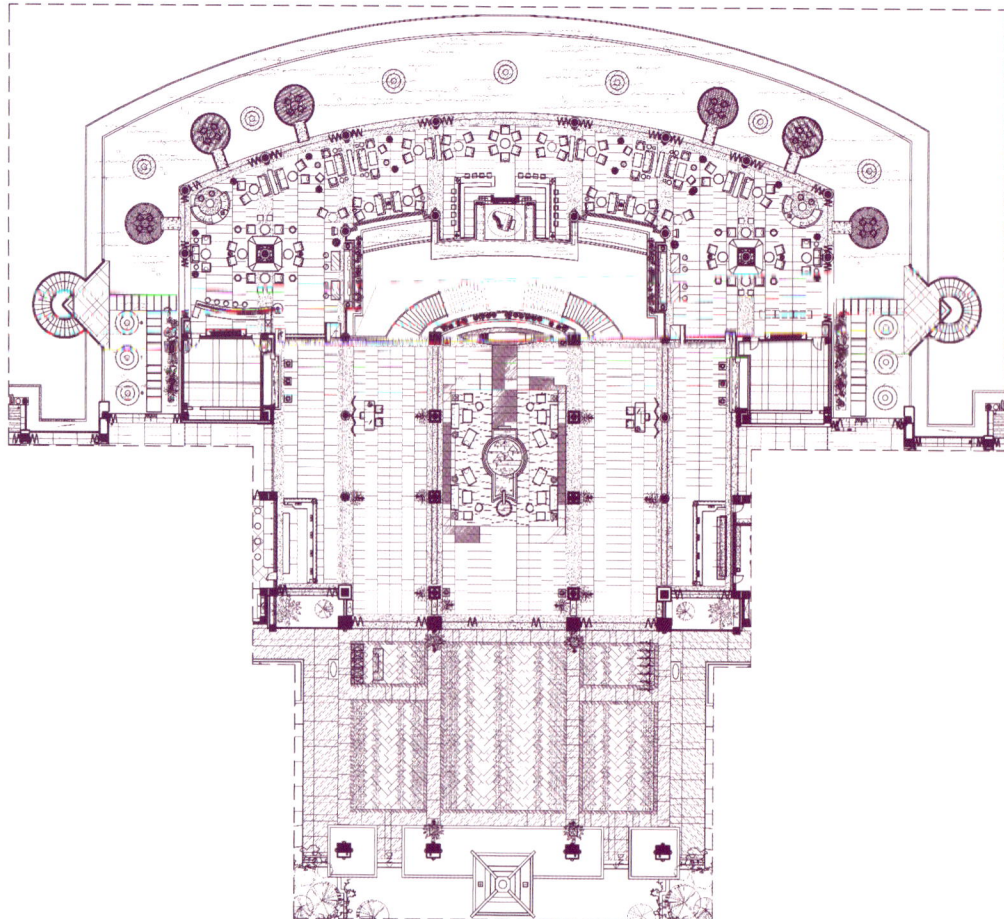

↑大堂休息区 / Lobby lounge area.
←错落有致的吊灯源于传统渔具造型 / Patchwork of chandelier style rooted in traditional gear.
↓大堂幕墙立面图 / The lobby wall elevation.

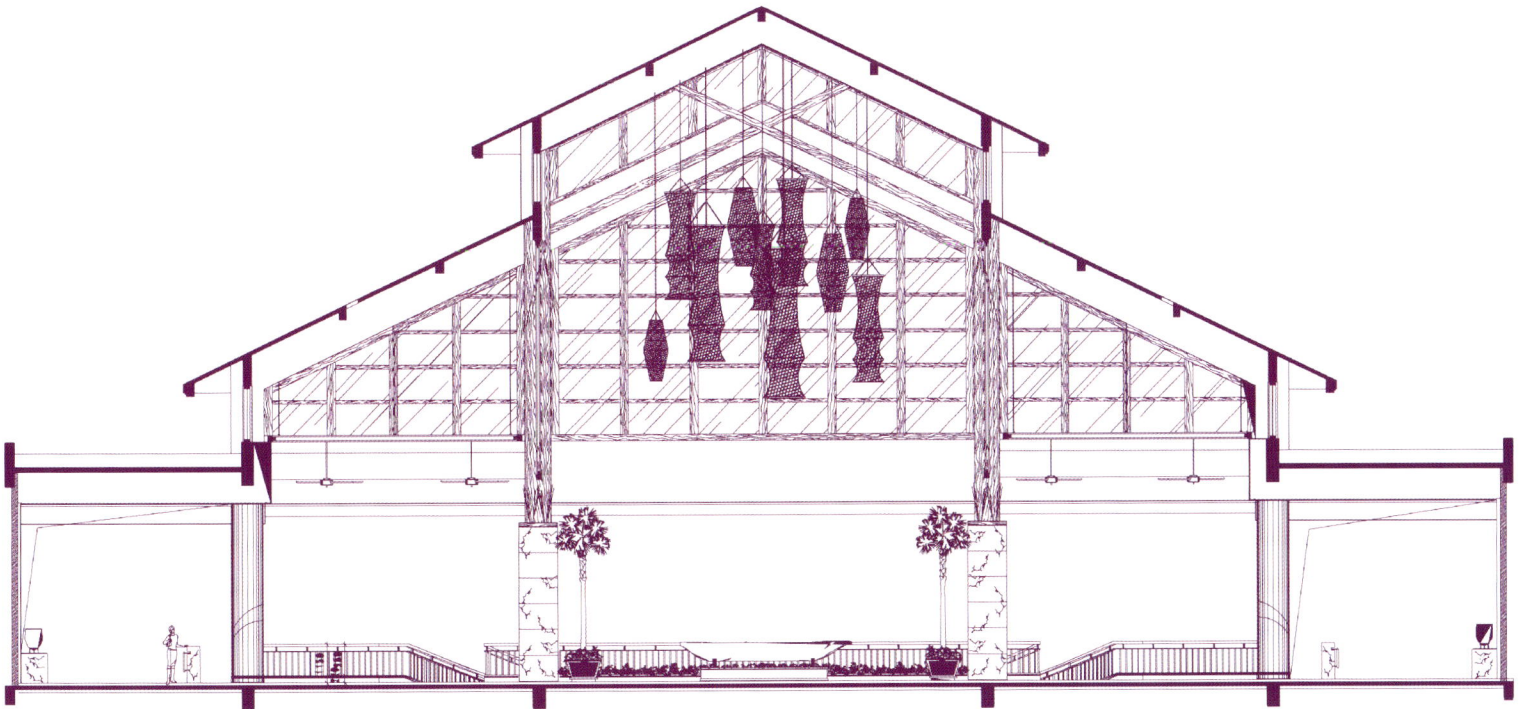

↑别墅卧房 / Bedroom villas.
←电梯厅延续了大堂的装饰风格 / Continuation of the lobby elevator hall decor.
↓标准客房 / Standard rooms.

ATLANTIS
IN DUBAI
棕榈岛亚特兰蒂斯酒店

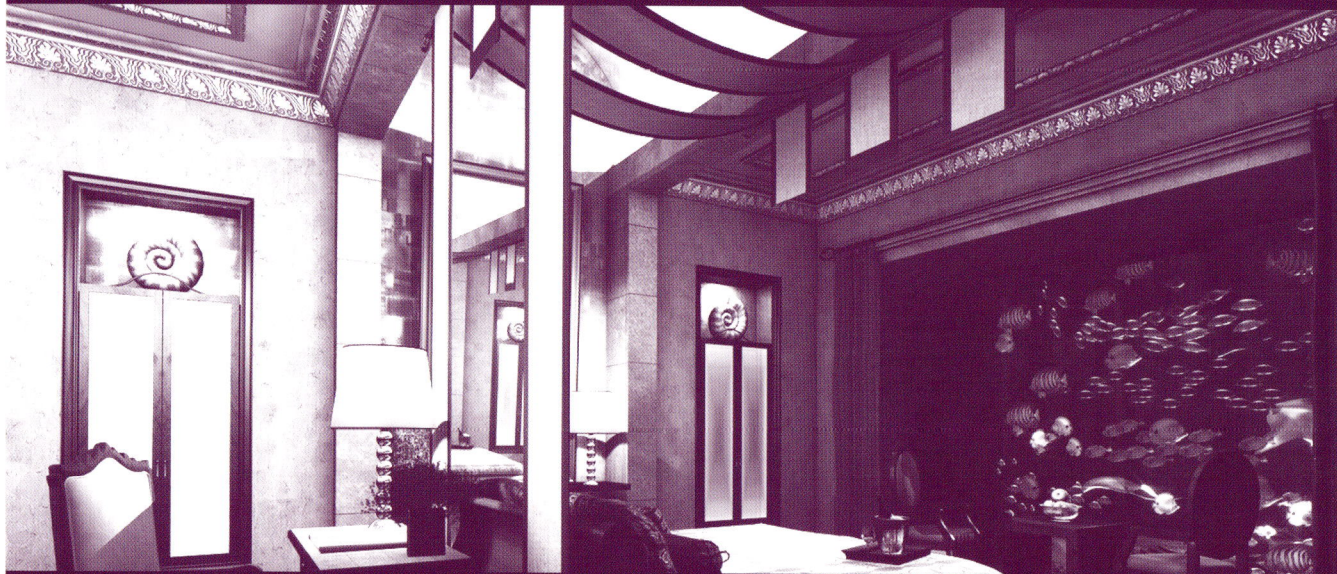

16

【坐落地点】迪拜棕榈岛；【造价】120亿元（RMB）
【工程名称】迪拜棕榈岛亚特兰蒂斯酒店
【设计】Wilson Associates〔美〕
【摄影】Courtesy of The Palm

从进入酒店大堂的那一刻开始，客人不由自主地眼花缭乱起来，以希腊神话中的众神为元素的壁画营造出富丽堂皇的精致感。8根鱼形柱撑起了一个海胆状的中庭，一座9 m高水晶雕塑矗立其中，造型极像海底的神秘生物，象征着亚特兰蒂斯众神力量的源泉。

亚特兰蒂斯酒店从一开始就宣布要"成为一个主打海洋主题的家庭娱乐胜地"。1539间客房的巨大建筑体量是其他同类酒店无法比肩的，尤其是位于酒店东西双塔连接通道处的"廊桥套房"堪称酒店客房中的极品，价格达每晚3.5万美元。除去拥有17家各国风味的餐厅、酒吧、一系列世界顶级名牌专卖店等豪华酒店所必不可少的设施，面积达20 000平方英尺的水疗馆绝对是惊喜中的惊喜。

一组大型的海底迷宫——消失的密室（Lost Chamber），是该酒店最大卖点所在。这座海底迷宫占据了酒店的两层楼，由不同的管道和宫殿连接组成，客人在寻找出口的同时，可以观赏游弋在身旁的65 000只全球各地海洋生物。

From entering the lobby of that moment, dazzled the guests could not help themselves, to the gods in Greek mythology to create a mural for the elements of an exquisite sense of grandeur. 8 column propped up a sea urchin, fish-shaped atrium, a 9 m high crystal sculpture, which stands, looks very like the mysterious undersea creatures, a symbol of the gods of Atlantis source of strength.

Atlantis hotel announced from the outset, "has become a main marine-themed family entertainment destination." 1539 rooms of the grand architecture of body volume is not par with other similar hotels, especially in East and West Towers hotel connection channel at the "Corridor Bridge Suite" rated hotel rooms in Need, the price reached 35,000 U.S. dollars per night. Remove has 17 international restaurants, bars, a series of brand-name stores such as the world's top luxury hotels of the essential facilities, an area of 20000 square-foot Spa is absolutely pleasant surprise in the surprise.

A group of large-scale undersea labyrinth - the disappearance of the Chamber of Secrets (Lost Chamber), is the biggest selling point of the hotel is located. This undersea maze occupied two floors of the hotel, from palaces to connect the different pipes and the composition of exports, while the guests are looking for, you can watch stayed in the side of 65 000 marine life around the world.

↑金碧辉煌的休闲区 / The magnificent recreation.
←酒店大厅 / Hotel lobby.
↓标准客房平面图 / Standard room layout.

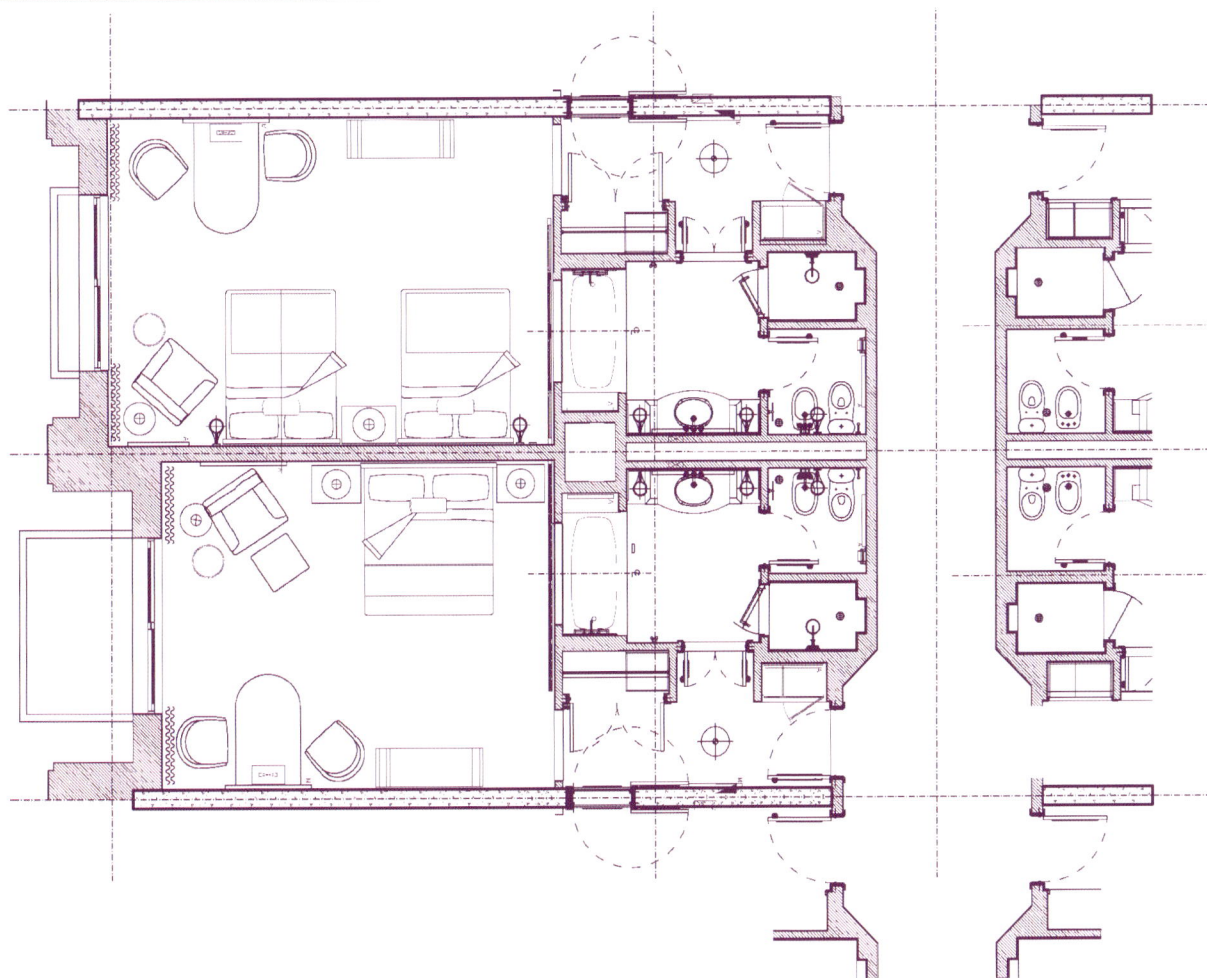

↑蓝色的海洋梦幻是酒店主题 / Blue Ocean Dream theme of the hotel.

↑酒吧 / Bar；↑尖顶的拱门，尤满阿拉伯风情 / Steeple of the arches, is full of Arab customs.
↓标准客房 / Standard rooms.

ART
HOT

艺术酒店
ELS

VISION
FASHION HOTEL
视界风尚酒店

17

【坐落地点】深圳深南大道深圳大剧院内；【面积】4300 m²
【工程名称】深圳视界风尚酒店
【设计】荣益
【摄影】贾方

酒店位于深圳市的最繁华路段，极佳的区位成就了它的高人气。设计师希望这里是展现深圳这座城市所代表的开放、创新、包容、现代感和国际化视野的最佳场所。

酒店接待大厅的云纹装饰吊顶是设计师的得意之作，泛着淡淡的金属光泽，一种静止的波诡云谲让人眼前一亮。整个大厅用色简单，仅有的黑白两色，通过不同的造型、各异的图案、深浅的颜色差异，将整个空间表现得层次感十足。人厅旁的餐厅也是如此，红白黑的主色调，配搭得恰到好处，时尚感十足。

在对房间充满期待的同时，酒店的公共空间也让每一位前来的人看到了设计的美好。从纯净的白色，到热烈的红色，再到诡秘的黑色，仅是色彩的变换就让顾客应接不暇。房间的设计近乎设计师的自我实现，天马行空的美好愿景在这里得到了终极表达，就此创造出一个喧嚣与静谧并存的小世界。

Hotel is located in Shenzhen, the most prosperous section of an excellent location the achievements of its high popularity. Designers want to show here is represented by the Shenzhen city open, innovative, inclusive, modern and international vision of the best place.

Moire hotel reception hall ceiling is decorated designer's finest creations, suffused with faint metallic luster, a static lumpy it all themselves. A simple use of color throughout the hall, the only black and white, through different forms, different patterns, shades of the color difference between the performance of the entire space was a sense of the full level. Hall next to the restaurant is also true of the primary colors red, white and black, match just right, the fashion sense of the full.

In the room filled with expectations at the same time, the hotel's public space so that every one who came to see the beautiful design. From the pure white, to warm red, to the secretive black, only color change let customers overwhelmed. The design of the room near the designer's self-realization, a bright future here has been the ultimate expression of this connection to create a buzz with the quiet co-existence of small world.

↑黑白色系的大堂 / Black and white color of the lobby.
→楼梯间的吊灯 / Stairwell chandelier.
↓大堂平面图 / Hall floor plan.

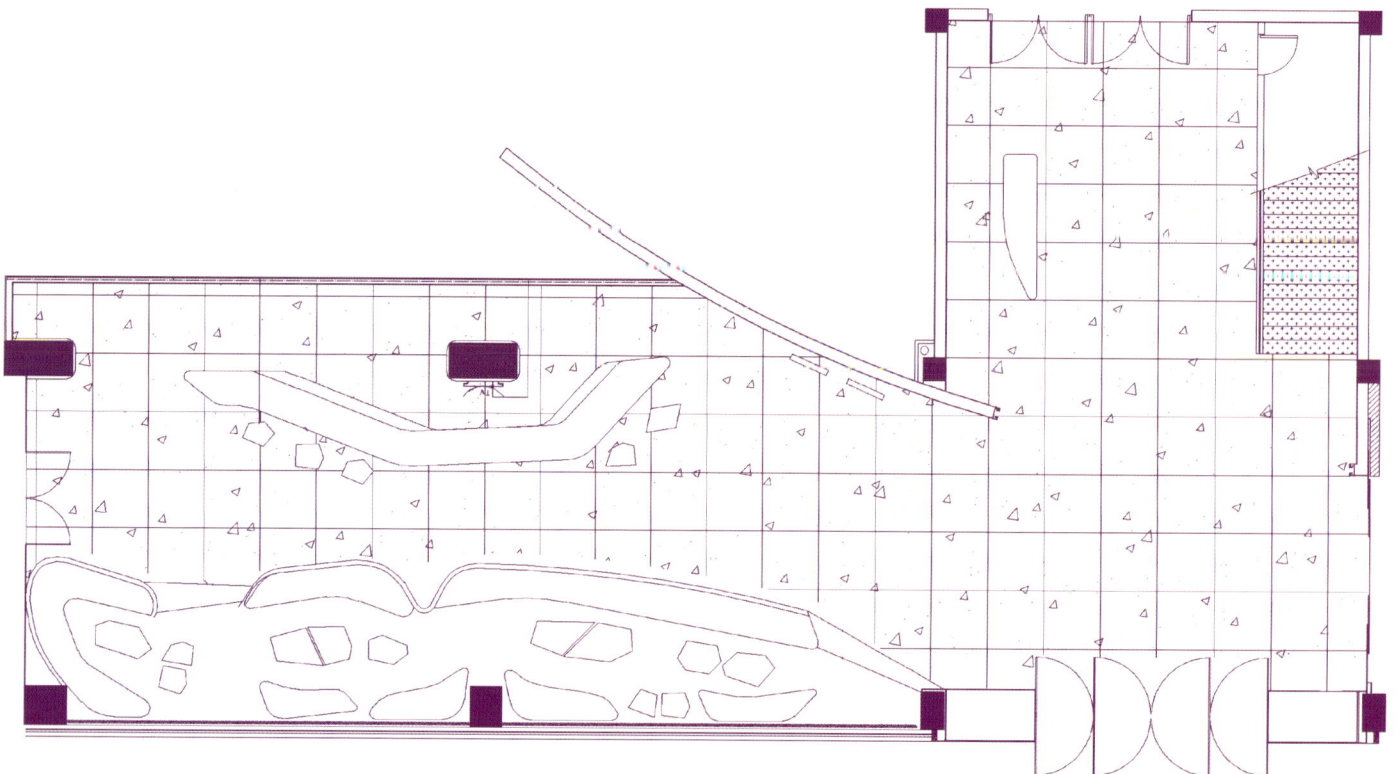

↑餐厅 / Restaurant. ↑红色的隔断增加了空间感 / Cut off the red increased sense of space.
←流线型的云纹吊顶 / Streamlined Moire ceiling.
↓标准客房平面图 / Standard room layout.

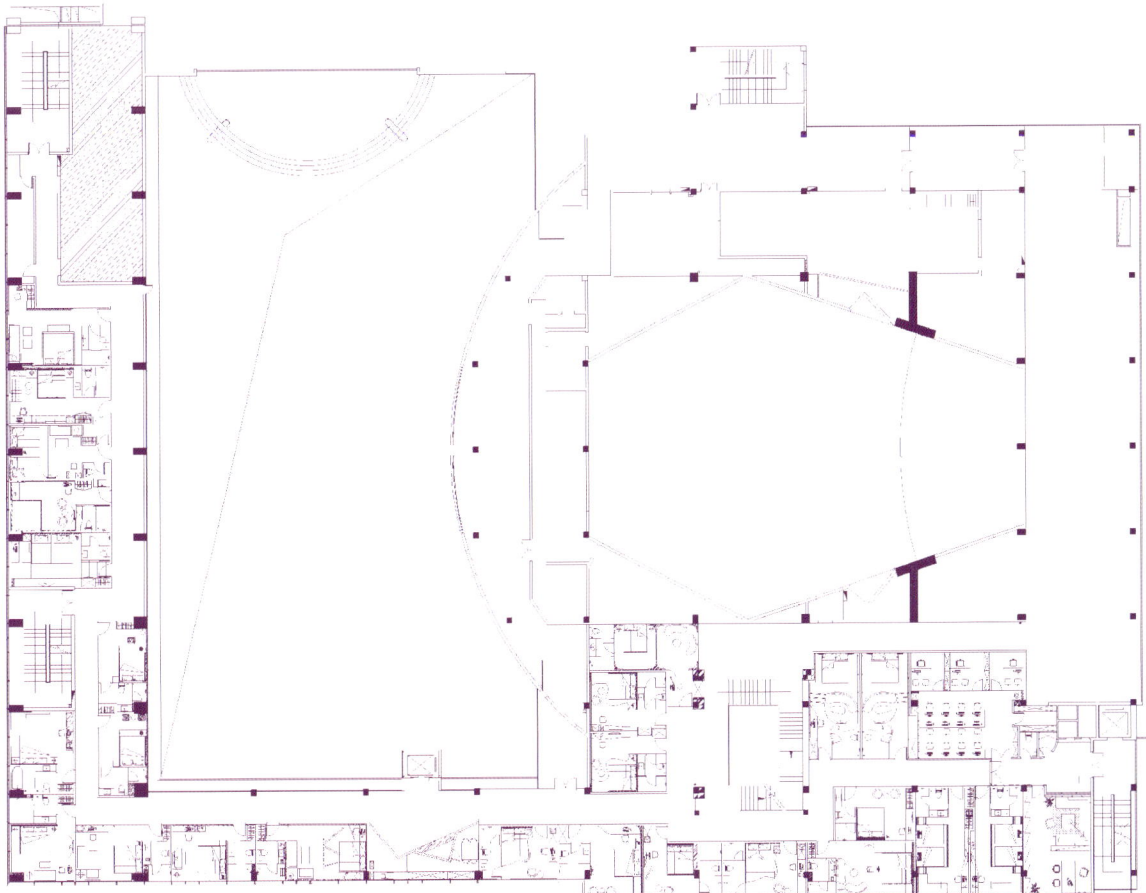

↑几何感十足的走廊 / Geometric sense of the full corridor.
←电梯间 / Elevator
↓洞穴造型走廊 / Cave shape corridor.

↑ "玻璃盒子"里的图案充满了奇妙的想像 / "Glass box" where the pattern is full of wonderful imagination.
→设计个性化的大套房 / Design personalized suites.
↓沉稳的用色配以时尚的图案 / Steady fashion coupled with the use of color patterns.

↑有着繁复图案的"黑色舞幔" / Has a complex pattern of "black dance mantle."
‹ 洞穴造型的房间 / Cave shape of the room.
↓大幅的手绘墙画让房间更有艺术感 / Substantial hand-painted wall art make the room more artistic.

↑极简的黑白两色即描摹出中东风情 / Minimalist black and white colors portray a Middle East style.
←木容舱造型的客房 / The capsule-shaped rooms
↓设计个性化的大套房 / Design personalized suites.

SEEKO'O
HOTEL

波尔多Seeko'o旅馆

18

【坐落地点】法国波尔多市夏特隆区；【面积】2300m²
【设计】Atelier d'architecture King Kong 〔法〕
【参与设计】Paul Marion、Jean-Christophe Masnada、Frederic Neau、Laurent Portejoie
【摄影】Arthur Péquin

从建筑到室内，整座旅馆呈现出雕塑般的观感，简洁的几何线条搭配建筑墙面上灵巧的转角、不规则的窗户，以及阴影、灯光等的设计，都为旅馆增添了现代动感。旅店的内部设计也是尽可能的自由简洁。

一走进旅馆，立刻能感受到不同于一般旅馆的惯常形象，映入眼帘的是充满普普风设计感的摆饰与家具。室内主色彩以白色为主调，辅以沉稳的黑色划分空间，并选择性地应用热烈的红色，或是大片洒泼，或是小笔点缀，打破黑白的沉闷。同样，旅店一层大厅内4.5 m长的接待台同样是由可丽耐做成。从外到内，总能看到可丽耐的踪影，而且形状都和简洁的建筑外形相似，时刻提醒着我们这座建筑的存在。

旅馆共有45间客房，从28 m²到55 m²不等，其中5间房间置于顶楼，房内附设观景台。躲藏在Seeko'o的玻璃窗后，欣赏着加伦河的无限美景，静静安享带有葡萄酒味道的甜美时光。另外，让人感到有趣的是房间浴室内的洗手台，外形设计成如同Seeko'o 外观般的白色几何立方体，而坐落的位置并非如传统般镶靠在墙边或是委屈地栖身浴室一角，而是落落大方地耸立在浴室中央，这样的设计可以方便让两个人从不同的方向同时使用一个洗手台，在有限的空间中做最有效的设计，这也是设计师考虑的重点之一。

From construction to interior, the entire hotel showing a sculpture-like look and feel, simple geometric lines with the corner wall of a building on a smart, irregular windows, and shadows and lighting design, both for the hotel added a modern dynamic. Hotel's interior design is simple as far as possible free.

Entering the hotel, you are immediately felt different from the usual image of the hotel in general, greets us is full of a sense of decoration and furniture design. Indoor white-based color tone, combined with steady black division of space, and selective application of warm red, or a large spill pouring, or small decorations, to break the boring black and white. Similarly, the hotel lobby within the 4.5 m-long reception station is also made from Corian. From outside to inside, always see the trace of Corian, and the shape and simplicity of the architectural features similar to, always reminds us of the building's existence.

A total of 45 hotel rooms, from 28 m² to 55 m² range, of which 5 rooms placed in the top floor, the room-cum-observation deck. After hiding in the Seeko'o windows to enjoy the infinite beauty of the Garonne, quietly enjoy time with the sweet taste of wine. In addition, people find it interesting that the room the bathroom hand-washing station, shape is designed to look like white as Seeko'o geometric cube, while the position is located is not as traditional as framed and leaning against the wall, or grumbled, shelter bathroom corner of but the graceful manner stands in the bathroom of the Central, so the design can be easily allow two people from different directions at the same time using a hand-washing station, in the limited space to do the most efficient design, which is considered one of the key designers.

↑ 酒店的建筑外观 / Architecture of the hotel exterior.

↑窗户的玻璃上印刻着形态各异的图案 / The glass windows inscribed with patterns printed on different patterns.
↓纯白的、不规则的建筑外立面 / Pure white, irregular building facade.

↑ Seeko'o旅馆全景 / Seeko'o Hotel Panorama.
↓ 平面图 / Plan

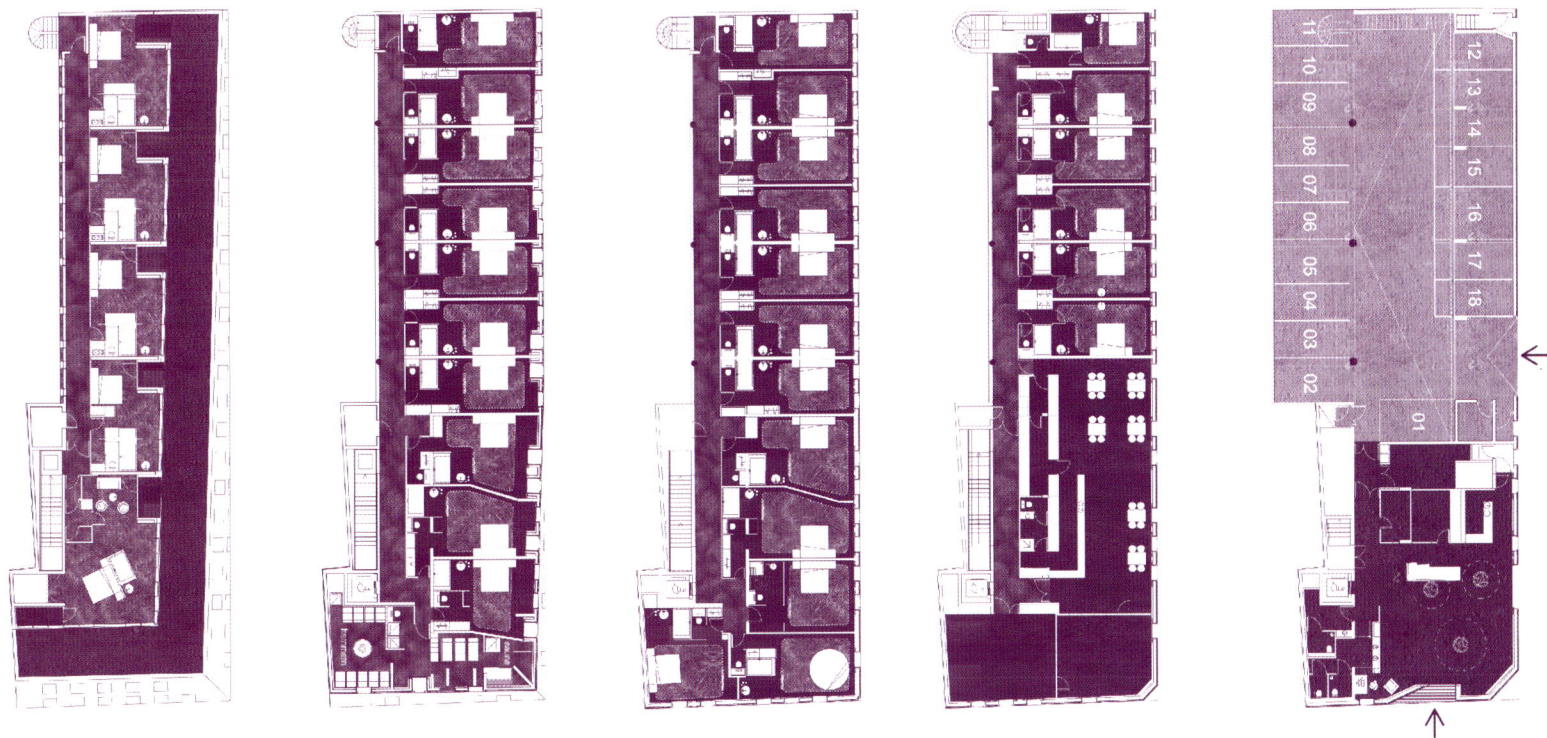

CHAMBRE/ROOM-101

↑↑旅馆的走廊 / The hotel corridor.
←顶层的豪华客房 / Top-level luxury guest rooms.
↓不规则的墙面映在镜面天花上，拉伸空间的高度 / Irregular walls reflected in the mirror on the ceiling, stretching space height.

LANCHID 19

H O T E L

Lanchid 19 设计酒店

19

【坐落地点】Lánchíd utca 19，布达佩斯，匈牙利
【规模】45间标准房，3间套房
【室内设计】Péter Sugár、Lázló Benczúr、Dóra Fónagy 〔匈〕
【建筑设计】Péter Sugár、Lázló Benczúr、Dóra Fónagy （D24）

酒店方十分注重艺术的创造性，注重建筑师与室内设计师的配合。作为设计酒店，Lanchid 19 Hotel的每一个房间都有各自的特色，都有不同的创作主题：有的以甜蜜的爱情为基调，有的则像一部经典电影的某个场景，有的是一堂舞蹈课。

整个酒店的设计思路来源于酒店所处的优美环境和这个古老城市的历史积累，也充分强调了游客们积极分享的重要性。酒店的建筑外观像一架玻璃的手风琴页片，通透的建筑本身就是一件艺术品。

酒店墙裙部分保留了中世纪建筑的特点。建筑前身是一个水塔，负责为Buda皇家城堡供水。在施工改造的时候并没有对这部分墙体做太大改动。酒店还保留了大厅的玻璃地面、多功能厅、葡萄酒品尝吧、展览厅等原有建筑的功能区域。

酒店创造性的设计理念来源于匈牙利一个新兴的联合设计体系，许多来自不同领域的艺术家（如平面设计师、摄影师、时装设计师等）聚集在一起用他们各自最熟悉的方式来共同解决不同文化背景的游客对酒店的需求。酒店共有45间风格各异的客房和3间套房可供选择。

Hotel places great emphasis on the creative arts, focusing on co-ordination of architects and interior designers. As a design hotel, Lanchid 19 Hotel Each room has its own characteristics, has a different creative theme: Some sweet love with the tone, while others, like a classic movie one scene, either a dance lesson .

Throughout the hotel's design ideas from hotel situated in a beautiful environment and the history of this ancient city the accumulation of fully stressed the importance of visitors to actively share. Architecture of the hotel looks like an accordion page piece of glass, transparent building itself is a work of art.

Some of the reservations of the hotel's walls medieval architecture features. Building was formerly a water tower, is responsible for the Buda Royal Castle water supply. Transformation of the time in the construction of this part of the wall do not change too much. The hotel also retained the hall of glass on the ground, multi-purpose hall, wine tasting bar, an exhibition hall and other functional areas of the original building.

Hotels creative design concepts from Hungary, a new co-design system, and many artists from different fields (such as graphic designers, photographers, fashion designers, etc.) together with their respective most familiar way to a common address different cultural background of tourists to the hotel needs. A total of 45 different hotel-style rooms and three suites to choose from.

↑酒店的建筑外观 / Architecture of the hotel exterior.

↑酒店的夜景 / The hotel's night.
→粗犷的石材墙面增加了餐厅的设计感 / Rugged stone walls increased the restaurant's design sense.
↓酒店入口 / Hotel entrance.

↑早餐吧 / Breakfast bar.
↓灯光下的早餐吧 / Under light breakfast bar.

↑客房大厅保留了原有的玻璃地面 / Room lobby retains the original glass surface.
↓休息区 / Rest area.

← 夸张的卫生间装饰 / Exaggerated bathroom decoration.
↓ 卫生间的对面是多瑙河 / The Danube opposite the bathroom.

XINGYUEWAN
H O T E L

星月湾时尚快捷酒店

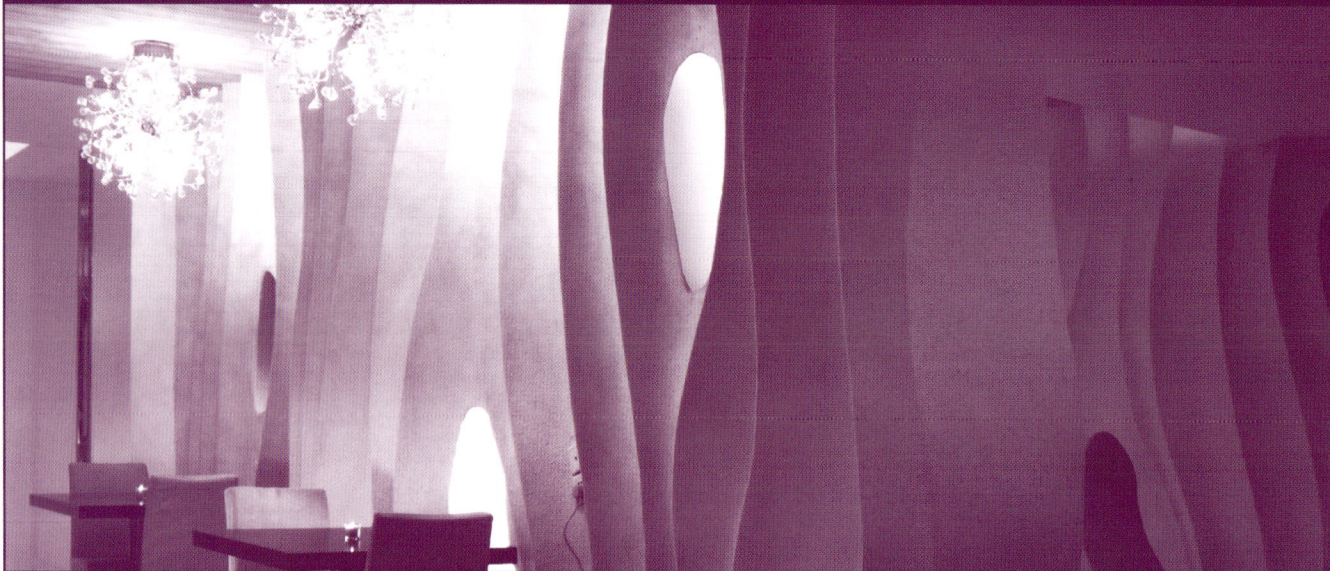

20

【坐落地点】长春市隆礼路；【面积】3500 m²
【设计】陈旭东
【主要材料】爵士白理石、玻璃钢、银镜
【摄影】陈旭东

酒店是由一栋旧住宅楼改造而成。经营方希望把大堂与餐厅设在一楼，但又不能改变原有建筑的基本结构。设计师结合星月酒店一贯的经营理念，以及注重带给入住者舒适睡眠的倡导，最后决定用"星、水、月"作为设计母题，这也正好契合酒店"星月湾"的店名。

在实际设计中，最制约设计师的是原来旧车库中的隔墙。在这里用"水"的元素与原结构中的实墙结合，墙面蜿蜒流转，如同水面被风吹起的涟漪，层层延伸，不仅给规矩的室内空间带来动感，也完全消除了普通砖墙呆板的感觉。墙面上的开洞顺应墙面"水流"的动势，如同湖面上的月影被风吹扯动的形态，是"水中月"在空间中的具体体现。

在色彩使用上，整体色调以白色与暗红色的泥土色为主。白色让人联想到皎洁的月光，配合柔和的灯光，能产生直达人心安宁、静谧的力量。砖红色——代表东北地区特有的"高天厚土"的环境，厚实的泥土也带给人温暖与踏实。月光与泥土之间，天与地之间，晶莹的水晶灯如同星光点点，点缀其中，给安适的空间带来些许跳跃、活泼的感觉。

Hotel is a renovated old residential buildings. The operator wants to lobby and restaurant on the first floor, but can not change the basic structure of the original building. Designer with hotel business philosophy has always been, as well as focus on bringing an inmate advocacy comfortable sleep and finally decided to use "The Star, water month" as a design motif, which also just fit hotel "Xingyue Bay" in its name.

In practical design, the most constraining is the original designer of the old garage wall. Here with "water" element of the original structure of the combination of solid walls, walls meandering flow, such as the date of the wind ripples the surface, layer upon layer extends not only to the rules of the interior space to bring dynamic, but also completely eliminates the Common brick stiff feeling. Comply with the open hole on the wall the wall, "water" and momentum, as the lake wind shadow of months Chedong form, is "in the water" in the space in the concrete embodiment.

The use of color, the overall tone with white and dark red in the main. White people think of bringing a moonlight, with soft lighting, can have direct access to the people peace, quiet strength. Brick-red - on behalf of the Northeast unique "high-days-thick soil" environment, thick mud has also given people a warm and practical. Between the moon and the earth, between heaven and earth, crystal chandelier as little starlight, embellishment which brings a bit of space to the well-being jumping, lively feel.

↑酒店入口处 / Hotel entrance.

↑餐厅的地势低于大堂 / The terrain below the lobby restaurant.
↓餐厅吧台 / Restaurant bar；↓换一个角度看餐厅 / Another point of view restaurant.

↑ "水"的元素与旧结构中的墙体相结合 / "Water" element in the wall with the original structure, combined.
↓一层平面图 / Floor plan.

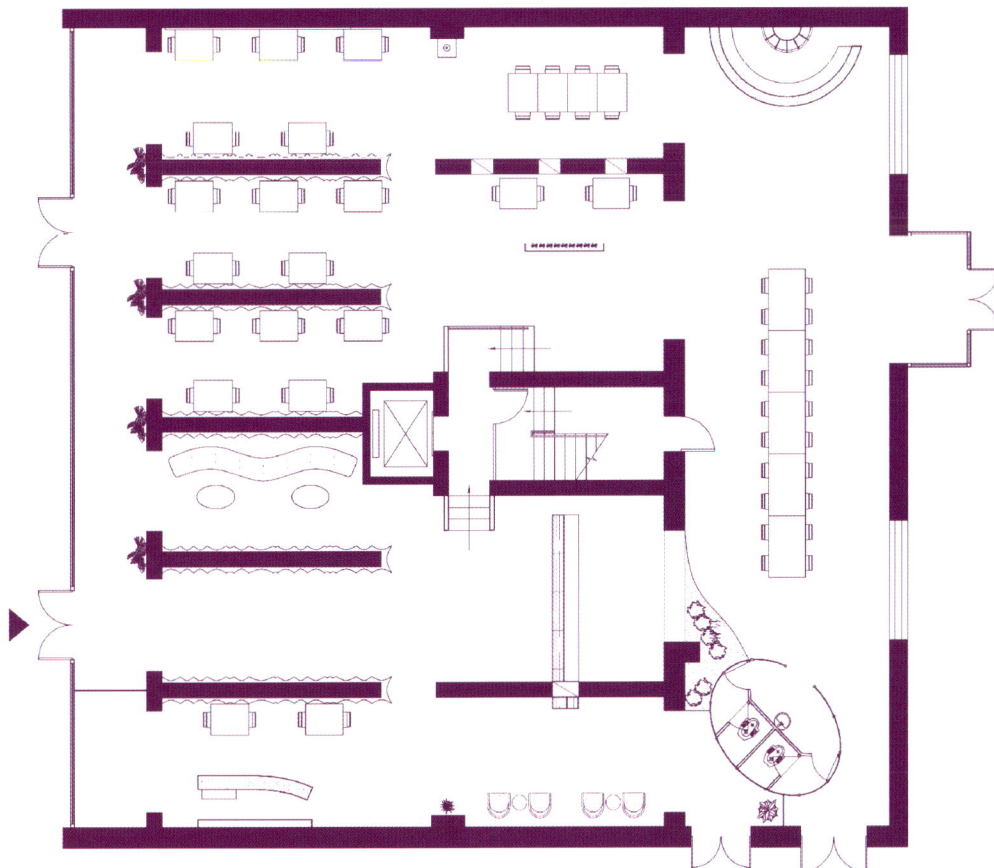

D U G E
COURTYARD BOUTIQUE HOTEL
杜革四合院精品酒店

21

【坐落地点】北京南锣鼓巷
【面积】606m²
【设计】Liu+de Biolley设计工作室
【摄影】申强、周之毅

杜革酒店是一个由四合院改建的艺术精品酒店，设计师让置身于其中的人体验到更多的建筑和空间之外对传统文化、历史的缅怀以及试图寻找一种恰到好处的新旧文化的糅合方式。

正规四合院一般依东西向的胡同而坐北朝南，基本形制是分居四面的北房（正房）、南房（倒座房）和东西厢房，四周再围以高墙形成四合，开一扇门。推门而入，原本的过廊被重新规划而有了一个更实际的功能——酒店的大堂。

穿过大堂就进入酒店的核心区域餐厅和客房区域。餐厅是酒店的活动和交流的中心，其平面位置也被规划在了四合院的中心位置，即庭院的中心，似乎从古至今这个中心位置从未改变过，改变的只是过往的世人。客房区域在空间的规划上是有难度的，把原有四合院的六间厢房改成十个客房，分别是牡丹亭、皇宫、西藏王国、竹屋、丝绸之路、金色池塘、白宫、野山、东方快车、激情岁月，是对中国印象的各个不同画面的浓缩，是中国文化的精髓。

DuGe Courtyard Boutique Hotel is a converted courtyard art boutique hotel, designer for exposure in which to experience more of the buildings and spaces outside of traditional culture, history, nostalgia and trying to find a just right combination of old and new culture .

Generally in accordance with the formal courtyard east-west north to south of the HuTong sit, the basic shape is separated on all sides of the hokubo (main building), the South Room (inverted seat room) and east-west wing. Pushed open the door into the room than the original gallery has been re-planning but also has a more practical function - the hotel lobby.

Entered through the lobby of the hotel's core areas of regional restaurants and guest rooms. Restaurant is the hotel's activities and exchange center, the planar location of the courtyard is also planned for the center, that is the center of the courtyard, it seems that ancient times to the center position has never changed, only the past to change the world. Room region in space planning is difficult, and the original courtyard of the six rooms into 10 rooms, namely, The Peony Pavilion, Royal Palace, the Tibetan kingdom, Takenoya, the Silk Road, Golden Pond, the White House, Wild, East Express, passion years, is a Chinese impression of the different images of the concentration, is the essence of Chinese culture.

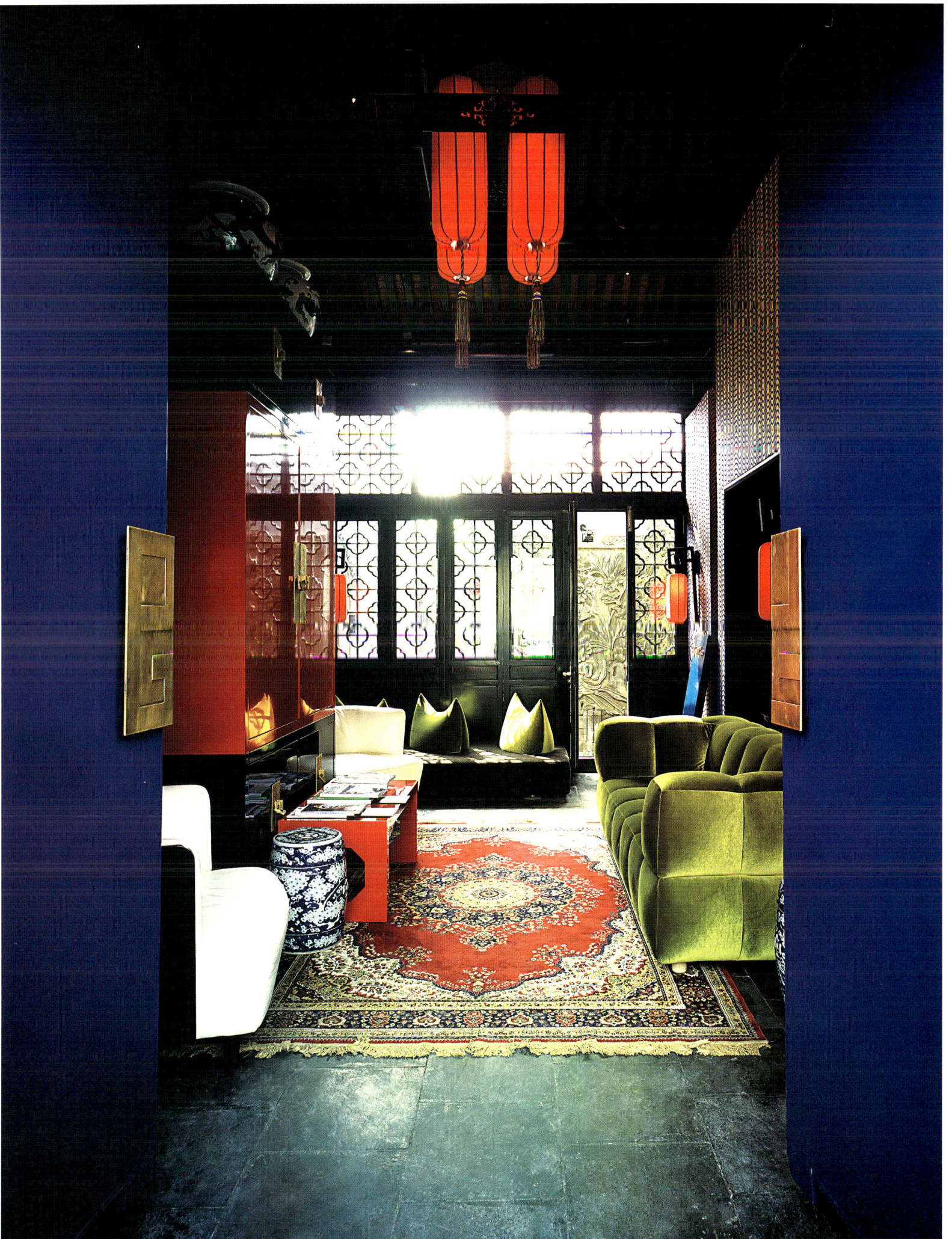

↑蓝色的大门里面是酒店的大堂 / The blue door inside the lobby of the hotel.

211

↑酒店中的亭子 / Hotels in the pavilion.

↑酒店中庭的小酒吧 / Hotel atrium bar.
↓平面图 / Plan

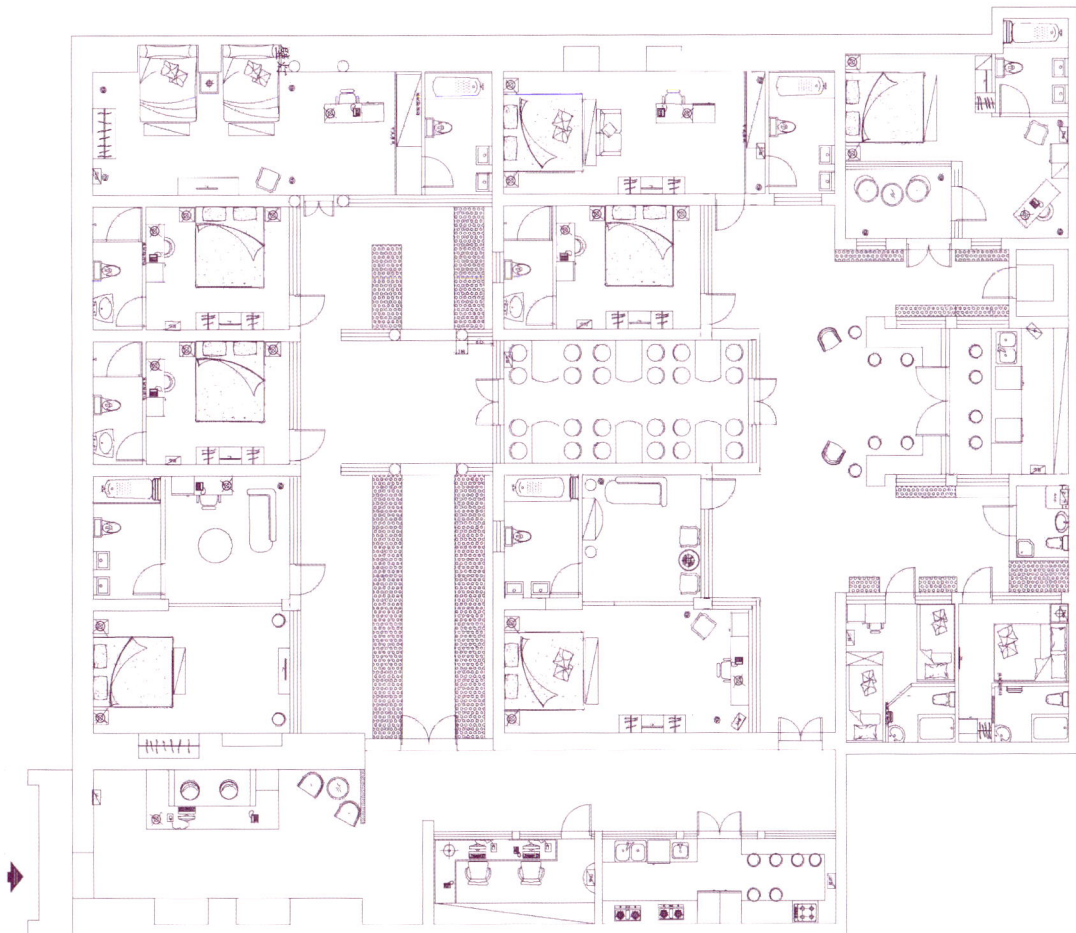

←↑ 整个院子的中心是酒店的餐厅 / The center of the courtyard is the hotel's restaurant.

↑ "金色池塘" 主题客房 / "On Golden Pond" theme rooms.
← "竹屋" 主题客房 / "Takenoya" theme rooms.
↓ "白宫" 主题客房 / "White House" theme rooms.

↑ "东方快车" 主题客房 / "Orient Express" theme rooms.
← "牡丹亭" 主题客房 / "Peony Pavilion" theme rooms.
↓ 客房的盥洗间 / Room toilets.

ANDEL'S
HOTEL LODZ

罗兹市安德尔酒店

22

【坐落地点】波兰罗兹市
【室内设计】Jestico + Whiles 〔英〕
【建筑设计】OP Architekten
【摄影】Ales Jungmann

本案设计师的使命就是找到一种全新的方式让建筑被感受和体验，在保留的原有工业结构与强烈节制的当代介入元素中找到平衡点。在整个功能布局上，四层挑空的中庭、时尚酒吧、咖啡馆、餐馆、商业中心以及大会议厅一应俱全，功能配套完善。

大堂上方空间并列着三个椭圆椎体的光井，由此构成中庭贯穿整个建筑物，这些倾斜的锥形向上延伸过，直达屋顶。餐厅位于首层，加上酒吧的现代设计，使它成为一个进行商业午餐和下午茶的好地方。

首层的会议室有着充足的日光，既可以打通成一个巨大的会议空间，又可将其分成7个各具风格的小会议室。四层的特别活动厅可容纳800人，并配置了空调和最新的视听设备。特殊的声学系统设计，让占地1300 m²的特别活动厅，可以承办各种活动，包括高品质的音乐会、发布会、展览，甚至是最新车型的官方展示。

设有标准卧房、长住公寓、无障碍客房以及拥有绝佳观景平台的总统套房。客房尺度合宜，施以明亮的色彩主题，高质量的家具和现代盥洗室，提供比同级别酒店更高品质的舒适标准。主要公共场所的品质和风格通过深褐色橡木家具和少量明亮的颜色贯彻到了所有卧房。

This case the designer's mission is to find a new way to make buildings were feeling and experience, while retaining the original industrial structure and the strong involvement of elements of contemporary restraint to find a balance point. In the whole functional layout, pick an empty four-story atrium, fashionable bars, cafes, restaurants, business center and the Conference Hall readily available, features a complete set.

Hall tied for the top of the three elliptical vertebral space of light wells, thus the composition of the Court throughout the building, which extends upward tilt of the cone had direct access to the roof. Restaurant is located in the ground floor, combined with modern design bar, making it a commercial good place for lunch and afternoon tea.

The first floor conference room with plenty of sunlight, either open up into a large meeting space, can also be divided into seven different styles of meeting rooms. Four of the special events hall can accommodate 800 people, and equipped with air conditioning and the latest audio-visual equipment. A special acoustic system design, so that an area of 1300 m² of special events hall, you can host a variety of activities, including high-quality concerts, conferences, exhibitions, and even the latest models of the official show.

With standard bedroom, long stay apartment, barrier-free guest rooms and has an excellent viewing platform for the presidential suite. Room scale appropriate to broach the theme in bright colors, high-quality furniture and modern bathrooms, providing higher quality than the same level of hotel standards of comfort. Major public places by the quality and style of dark brown oak furniture and a small amount of bright colors carried out in all the bedroom.

↑椭圆形的中庭直达屋顶 / Oval-shaped atrium directly to the roof; ↑ 变幻的色彩 / Changes color
← 大堂接待台 / Lobby Reception Desk.
↓ 一层和二层的平面图 / The first floor and second floor plan

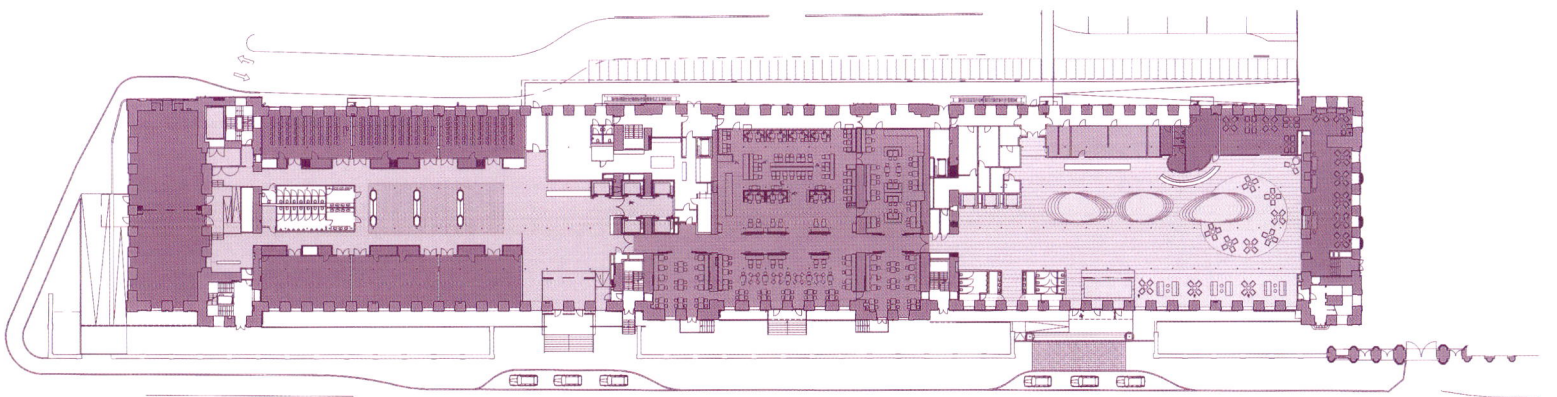

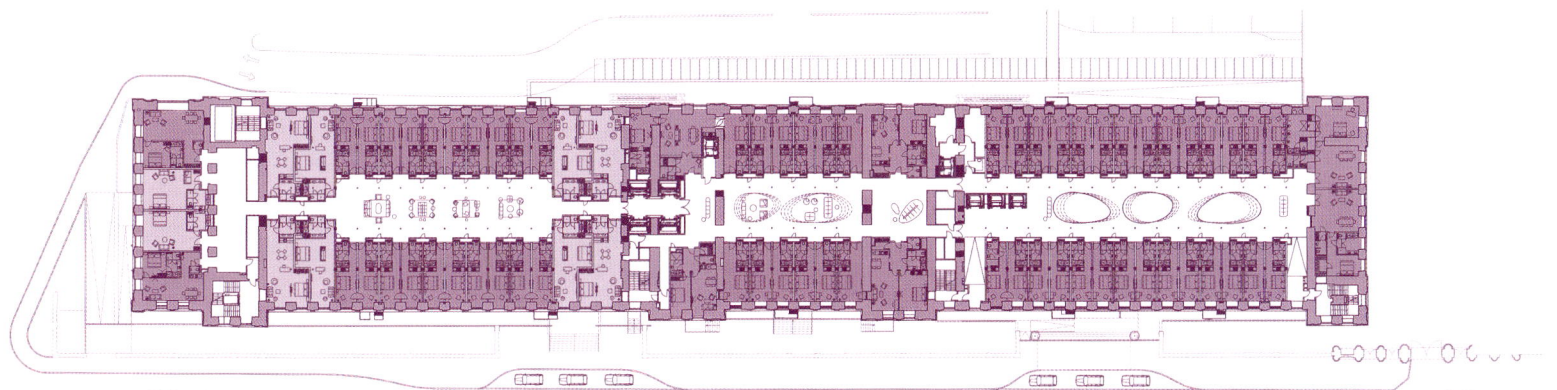

↑ 酒店的艺术墙 / Hotel art wall.
↓ 剖面图 / Section

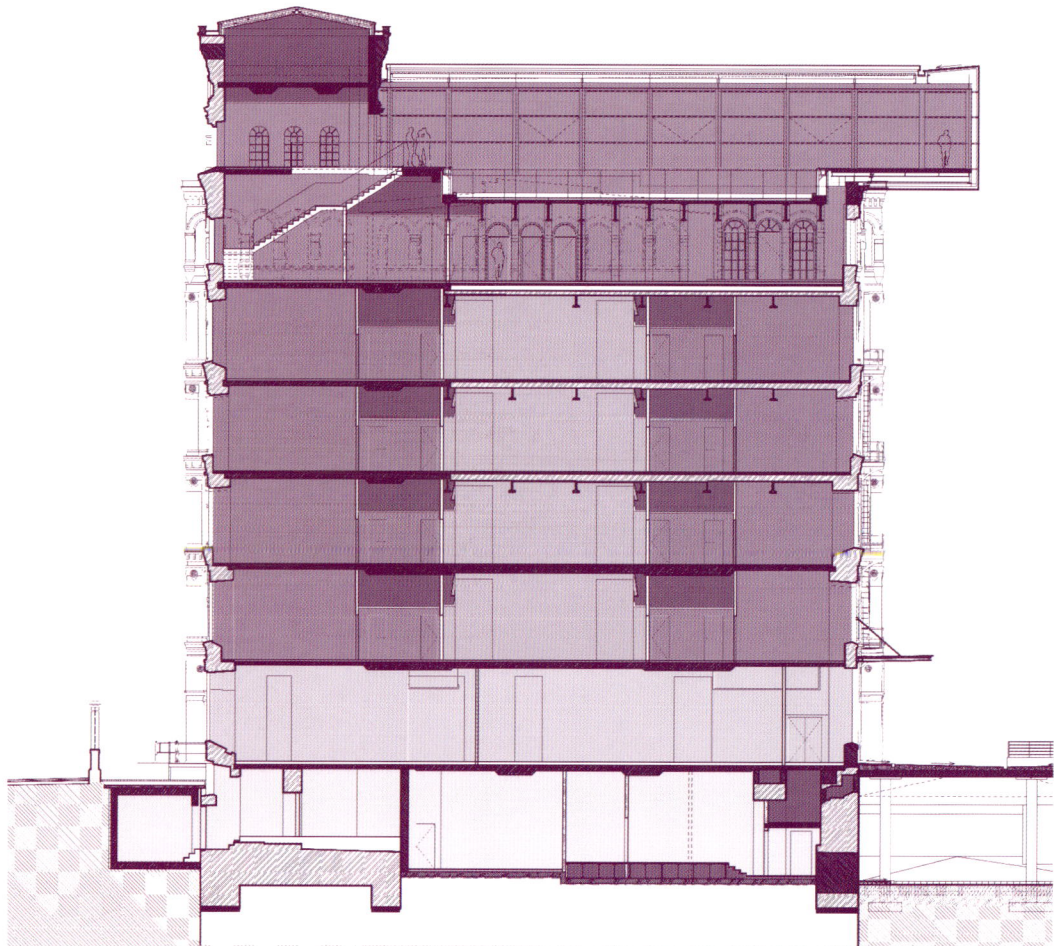

↑ 餐厅区 / Restaurant area.
↓ 吧台 / Bar

↑餐厅区原有的建筑结构格外明显 / Dining area the original architectural structure, particularly evident.
→裸露的红墙与现代感十足的餐厅 / Bare red walls and modern restaurant full.
↓另一风格的餐厅区 / Another style of restaurant area.

↑客房的会客厅 / Rooms of the living room.
←餐厅 / Restaurant
↓标准客房 / Standard room.

HOTEL G
AN EXQUISITE DESIGN CASE
精品设计酒店Hotel G

23

【坐落地点】北京朝阳区工体西路甲7号
【工程名称】Hotel G
【设计】Mark Lintott　Imaad Rahmouni〔英〕
【摄影】舒赫

在众多时尚餐厅和酒吧聚集的三里屯，白天的Hotel G以深紫色体块显得个性而稳重；而当夜色袭来，整齐的客房凸窗在LED灯的映照下从深色背景中跳跃出来，使酒店立即切换到动感、梦幻的夜生活氛围中。

客房的设计突出了灯光和颜色的搭配。紫色和咖啡色的混搭给人神秘、富贵和复古的感觉。暖色的灯光经过精致水晶珠帘的过滤，给暗色背景印上闪烁迷离的光斑，而铺着丝质床单的大床则让你平日超载的身心有了一种回家般的轻松。

为让多次来访的客人有不同的体验，酒店的110间客房以不同的设计分为4种类别：Good房通过蓝色格子背景营造了丰富的想象氛围；Great房优雅的茄紫色墙面和巨大的好莱坞黑白剧照让你恍如隔世；Greater就更为大胆，直接将绚烂的红色用于整个隔间，甚至让你有即将踏上星光大道般的兴奋；Greatest则以80 m²的奢侈面积和巨大的镜面,打造了一场亦幻亦真的视觉盛宴。

在餐饮娱乐方面，Hotel G也有不凡表现。酒店内有3个独特出众的餐厅：位于一层的顶级精致汉堡餐吧25 Degrees、位于8楼和10楼露台的时尚顶餐吧Gilt，以及高雅细致的日本料理Morio。

Among the many trendy restaurants and bars of Sanlitun gathered during the day of the Hotel G in order to deep purple body mass appears to personality and prudent; and when the darkness hit, clean room bay window in the LED light shine, from the jumping out of a dark background, so that Hotel immediately switch to the dynamic, fantastic nightlife atmosphere.

Rooms designed to highlight the mix of light and color. Purple and brown mashup gives mysterious, rich, and retro feel. Warm light filtering through the delicate crystal bead curtain, to be printed on dark background are blurred flashing spot, while the big bed covered with silk sheets then you have a normal physical and mental overloading as easy to go home.

To allow multiple visitors, have a different experience, the hotel of 110 rooms divided into 4 different categories of designs: Good room by a blue grid background to create a rich atmosphere of imagination; Great room and elegant eggplant purple walls and a huge black and white stills of Hollywood lets you Manasarovar; Greater even more boldly, directly to the gorgeous red for the entire compartment, and even give you embark on the Avenue of Stars-like excitement; Greatest luxury with 80 m² area and a huge mirror , to create a true visual feast is also magic.

In the dining and entertainment areas, Hotel G is also a remarkable performance. The hotel has a unique three outstanding restaurants: located in the top layer of fine hamburger bar 25 Degrees, located on 8th floor and 10th floor roof terrace of the trendy bar Gilt, as well as elegant and meticulous Japanese Morio.

←↑标准客房 / Standard rooms.
↓平面图 / Plan

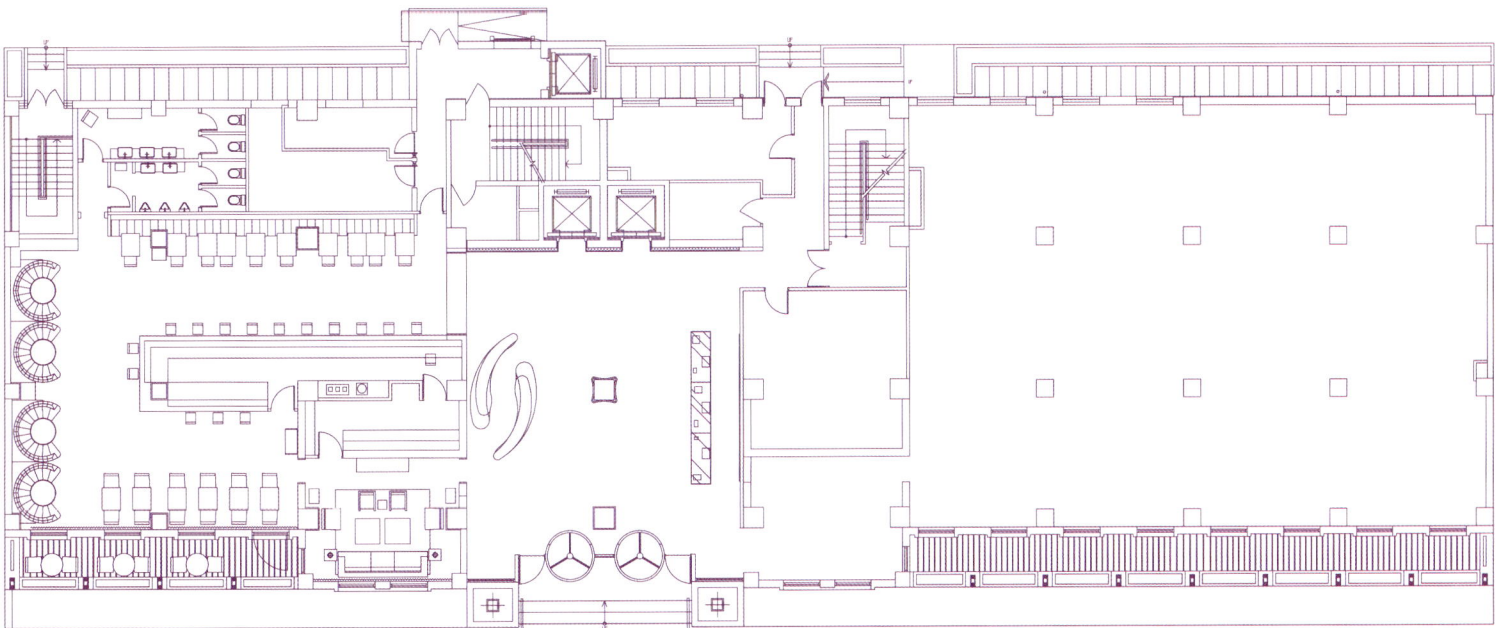

234

↑色彩和灯光划分了房间的不同区域 / Color and light into the room to a different area; ↑中式风格的博古架 / Chinese-style shelf.
→↓Greater房中热情的红色墙面和黑白海报充满了好莱坞时尚感 /
Greater room warm red walls and black and white poster is full of Hollywood fashion sense.

↑Murano长吧台加强了空间的纵深感 / Murano long space bar to enhance the depth.

（圆形的就餐桌 / Round the table.

↓丝绒图案的红色墙纸和暖色灯光让餐厅显得高雅而温馨 / Red wallpaper and warm lighting to the restaurant looked elegant and warm.

A TANG
HOTEL
阿汤楼酒店

24

【坐落地点】杭州山沟沟景区；【面积】3500 m²
【设计】周伟
【参与设计】陈于蓝、汪侠剑、徐寅莹、杨妍、杨小青
【摄影】贾方

酒店建筑为简洁的现代主义风格，灰色的清水砖墙配以黑色框架的玻璃幕墙，在密密的竹林前面简单的形体反而显得更加富有内涵。设计的最初动力源自对传统文化的现代诠释，沧桑的灰砖和围合的封闭性定义出外简内丰的朴拙性格，晶莹剔透的玻璃盒子和流动的空间给建筑带来了灵性与变化。传统的元素与现代元素的对峙与并存，既传达出对东方文化的尊重，又散发出一种现代的优雅气质。

走进酒店，首先进入接待区，接待区位于主入口的左侧，整个接待中心为黄色立方体空间，干净质朴，完整的几何形象使接待空间在整个建筑中的地位清晰地跳脱出来，接待中心与外部空间有堵透明的玻璃墙，设计者把由钢框架的玻璃盒子所围合的空间插入接待空间内，塑造了严谨而丰富的多层次空间：自然的外部空间、玻璃盒子的休憩空间、接待区的办公空间。三层建筑的主调分别是红、绿、黄，轻快的颜色和简洁的设计给人一种放松的感觉。

回到了客房，落地的大玻璃窗映入斑驳的竹影，帮游客控制思绪的节奏。目光从窗外移入室内，简单大气的墙绘定义着每个房间的氛围，50多个房间各有自己的特色，或宁静，或张扬，或诗意，总会有那么几个适合自己独特的心绪。

The hotel's architecture is simple modernist style, gray water from the brick wall with black frame glass curtain wall, in the thick bamboo forest in front of a simple shape but even more rich content. Design of the initial impetus comes from a modern interpretation of traditional culture, vicissitudes of gray bricks and enclosed by the closed nature of the definition of going out within the abundance of simple, innocent and simple character, sparkling glass boxes and mobile space for the construction and changes brought about spirituality . Elements of traditional and modern elements of confrontation and coexistence, both convey the right of respect for Eastern culture, but also exudes a modern elegance temperament.

Into the hotel, first enter the reception area, reception area is located in the main entrance of the left side of the reception center for the yellow cube space, clean and simple, complete, so that the geometric image of the reception space in the building's position clearly escape out of the reception center with the external space blocking transparent glass walls, the designer into a glass box by the steel frame of the space enclosed by inserting reception space, create a strict and rich multi-level space: a natural outer space, the glass box open space, the reception area of the office space. The main theme of the three-tier architecture are red, green, yellow, light colors and simple design gives a relaxed feeling.

Back to the room, floor large windows greet the mottled shadow, to help visitors controlled the rhythm of thoughts. Gaze shifted from the window room, the walls painted a simple atmosphere, the definition of the atmosphere of each room, more than 50 rooms have their own characteristics, or quiet, or quietly, or poetry, there will always be a few for their own unique frame of mind.

↑接待处的鹅黄色与灰砖在视觉上形成反差 / The goose that lays the reception yellow and gray bricks to form in the visual contrast.

239

240

↑大面积的暖色将整个空间打造得分外温馨 / Large areas of warm color to the entire space to create scoring warm outside.
→楼梯的围栏也精心设计过 / Stairs fence has been carefully designed.
↓↓长长的楼梯配合明快的颜色，让人心情愉悦 / Long flight of stairs with bright colors, people feel good.

↑不同的墙绘映衬出游客不同的心情 / Different wall painting gives a different mood.
↓一层平面图 / 1st floor plan.

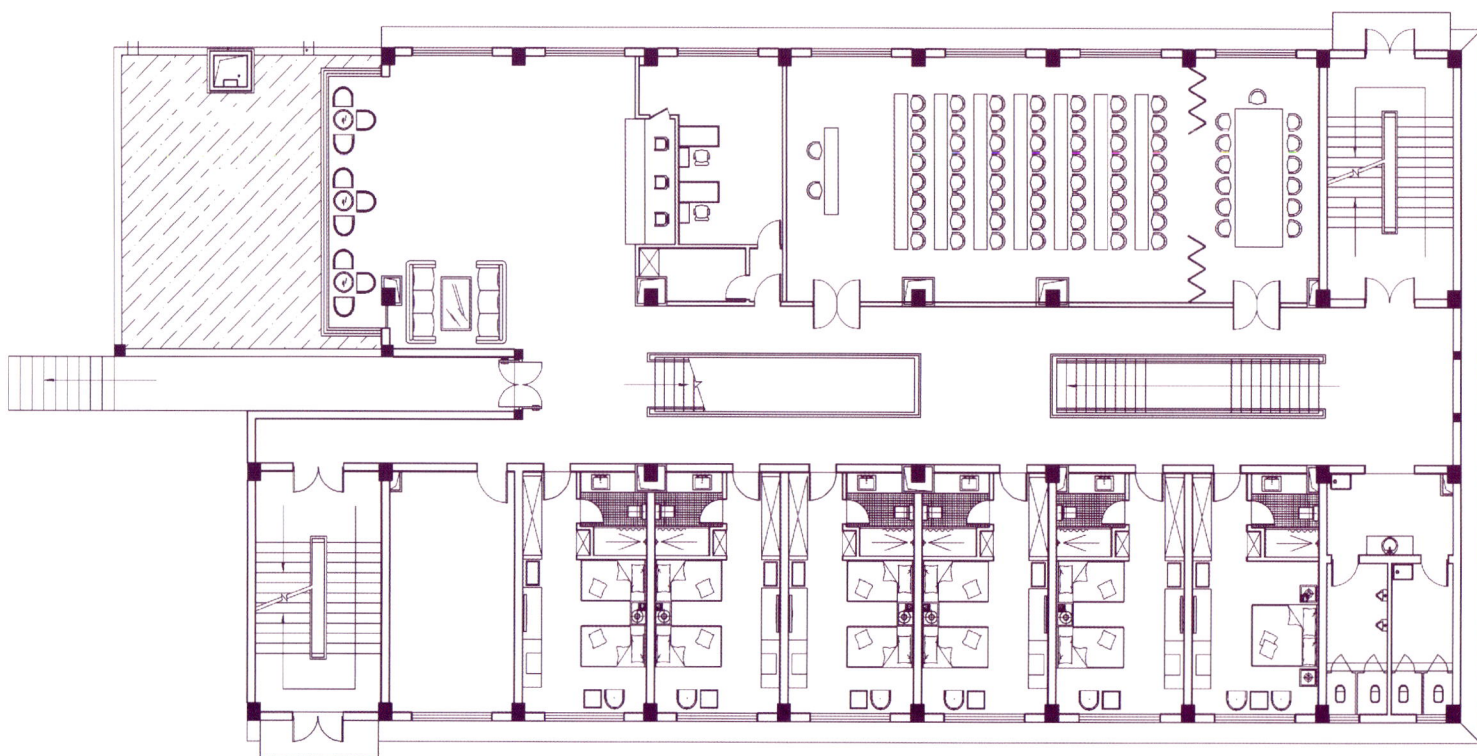

↑不同的墙绘映衬出游客不同的心情 / Different wall painting gives a different mood.
↓二层平面图 / 2nd floor plan.

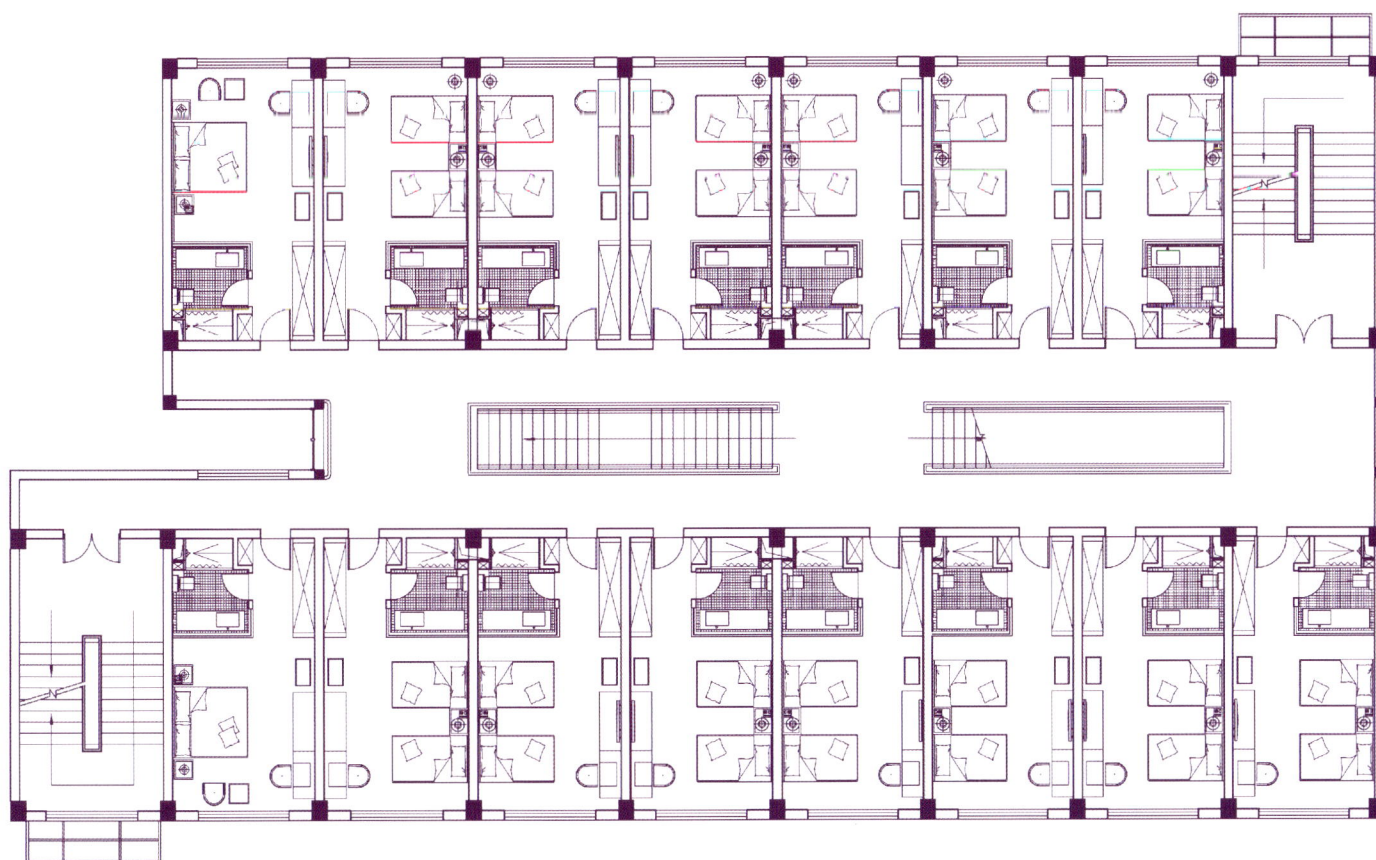

THREE
SISTERS' HOTEL
塔林三姐妹酒店

25

【坐落地点】爱沙尼亚塔林
【建筑设计】Martinus Schuurman 〔爱〕
【室内设计】Külli Salum 〔爱〕
【摄影】Courtesy of Design Hotels

这三栋房屋的整体外观都统一展现着建筑师自然质朴的设计风格，而室内设计师的首要任务则是力求维持三栋房屋内部相对独立、各具特色的装潢风格。

根据设计师的想法，拥有5间客房的 "小妹"，无论是墙壁上的摄影作品，还是新旧风格的强烈对比，都折射出她兼收并蓄的品位及对艺术的热爱；拥有7间客房的 "二姐" 装扮最为典雅别致，其中所有精美高贵的陈设都是从爱沙尼亚各处搜集而来的古董家具；"大姐" 共有8间稍小的客房，却处处散发着更加生动的气息，频繁举办的派对和酒会仿佛在向远道而来的朋友和访客们诉说她热情好客的生活方式。

在客房的设计上，每栋楼都有其鲜明的主题。"大姐" 的客房布置更趋于传统，由升降电梯可到达上部两层楼面；"二姐" 则提供独具特色的大型公寓式套房，配有独立的起居室和卧室；"小妹" 则分为上下两层，从楼下的客房可直接进入户外庭院，楼上则是起居室和主卧室组成的套间。

This is the overall appearance of all three houses to show unity with the natural simplicity of the design architect, while the interior designer's primary task is to strive to maintain the three houses within the relatively independent, distinctive decorating style.

According to the designer's ideas, has five rooms of the "young girl", whether it is photographs on the wall, or the sharp contrast between the old and new styles are a reflection of her eclectic tastes and passion for art; has seven rooms of the "second sister" the most elegant and chic dress, in which all the elegance of the furnishings are beautiful throughout the collection comes from Estonia antique furniture;" big sister "in a total of eight smaller rooms, but everywhere exudes a more lively atmosphere, often organized by parties and receptions as if coming from afar to friends and visitors have to tell her hospitable way of life.

In the room design, each building has its own distinct theme. "Big Sister" and tend to be more traditionally furnished guest rooms, from elevators to reach the upper two floors; "second sister" provides a unique large-scale apartment-style suites with separate living room and bedroom; "young girl" is divided into two from top to bottom, from the downstairs rooms can be directly into the outdoor courtyard, upstairs is composed of living room and master bedroom suite.

↑ 在古老的街道上可见酒店的外观 / Can be seen on the streets in the old hotel's appearance.

245

↑ 酒店的建筑外观 / Architecture of the hotel exterior.
↓ "大姐" 的酒店入口 / "Sister" of the hotel entrance.

↑ "二姐"套房中的牛皮制休闲躺椅 / Sister" system of leisure suite in the leather chair.
↓↓ "二姐"客房的其他部位 / Other parts of room.

↑ "小妹"中有分为上下两层的客房 / Rooms are divided into upper and lower two floors.
↓楼上的房间 / Upstairs room.

↑ "小妹" 中带有钢琴的套房 / Hotel suite with a piano
↓ "小妹" 的套房 / Hotel suites.

249

NHOW HOTEL

NHOW 酒店

26

【坐落地点】意大利米兰
【设计】Matteo Thun & Partners（意）； 【参与设计】Matteo Thun、Daniele Beretta
【配套设施】256 间客房（74间标准间，135间豪华套房，23间套房，1间总统套房，3个Spa馆）
【摄影】Courtesy of Matteo Thun & Partners

酒店的设计构思来源于"动力集装箱"的概念。设计师Matteo Thun在解释自己的设计时，用了"Design Fluid"这个词语。另外，酒店还和米兰的一些艺术沙龙、画廊、双年展长期合作，不定期地在酒店举行暂时性的艺术展览。

一进门就是一堵巨大的水泥墙，好像要在这城市中开始一段钢筋混凝土的冒险一样。酒店的灯光基本采间接照明，可以根据客人的要求制造出不同的氛围。

酒店的大堂完整地呈现了"Design Fluid"的设计理念，形式和功能的变化和转换展现了文化的多样性。类似折衷主义的处理手法，把各种元素糅合在一起，工业化进程的遗留物，现代艺术风格的装饰还有带有古典情味的家具和摆设。

Vip区的休息区，三盏吊灯均是出自Jacopo Foggini之手的名品——"花"。而餐厅入口处的超大尺寸的灯具也是出身名门，那是Catellani & Smith的设计作品。酒店的其他家具大多由Matteo Thun先生自己设计，看似平淡的家具在艳丽的背景衬托下显得更加高雅。白色树脂漆的家具和工业建筑给人的粗犷感觉完全不同，是那样的安静和优雅。

NHOW hotel's design concept derived from the "power container" concept. Designer Matteo Thun in the interpretation of his own design, using the "Design Fluid" word. In addition, the hotel and Milan long-term cooperation of some art exhibitions are often held in hotels a temporary art exhibitions.

A door is a huge wall of concrete wall, seemed to be in the city to start a venture like reinforced concrete. Hotel lighting basic mining indirect lighting, you can request of the guests to create a different atmosphere.

The hotel lobby complete presents "Design Fluid" design concepts, form and function to show change and transformation of cultural diversity. Similar to the eclectic approach, blends together the various elements, remnants of the process of industrialization, modern art styles are decorated with classic charm of the furniture and furnishings.

Vip area rest area, three chandeliers are from the hands of Jacopo Foggini Famous - "flower." The restaurant at the entrance of the large size of the lamps is also a well-born, it is Catellani & Smith's designs. Most hotels and other furniture designed by Matteo Thun husband himself, it seems plain furniture in the background of gorgeous backdrop of even more elegant. White resin, lacquer furniture and industrial buildings gives the rough feel completely different, are so quiet and elegant.

252

↑酒店入口处 / Hotel entrance；↑酒店接待台 / Hotel Reception Desk.
↓槽钢被保留在走廊的中间 / Steel frame has been retained in the middle of the corridor；↓休息区 / Rest area.

↑餐厅 / Restaurant
↓平面图 / Plan

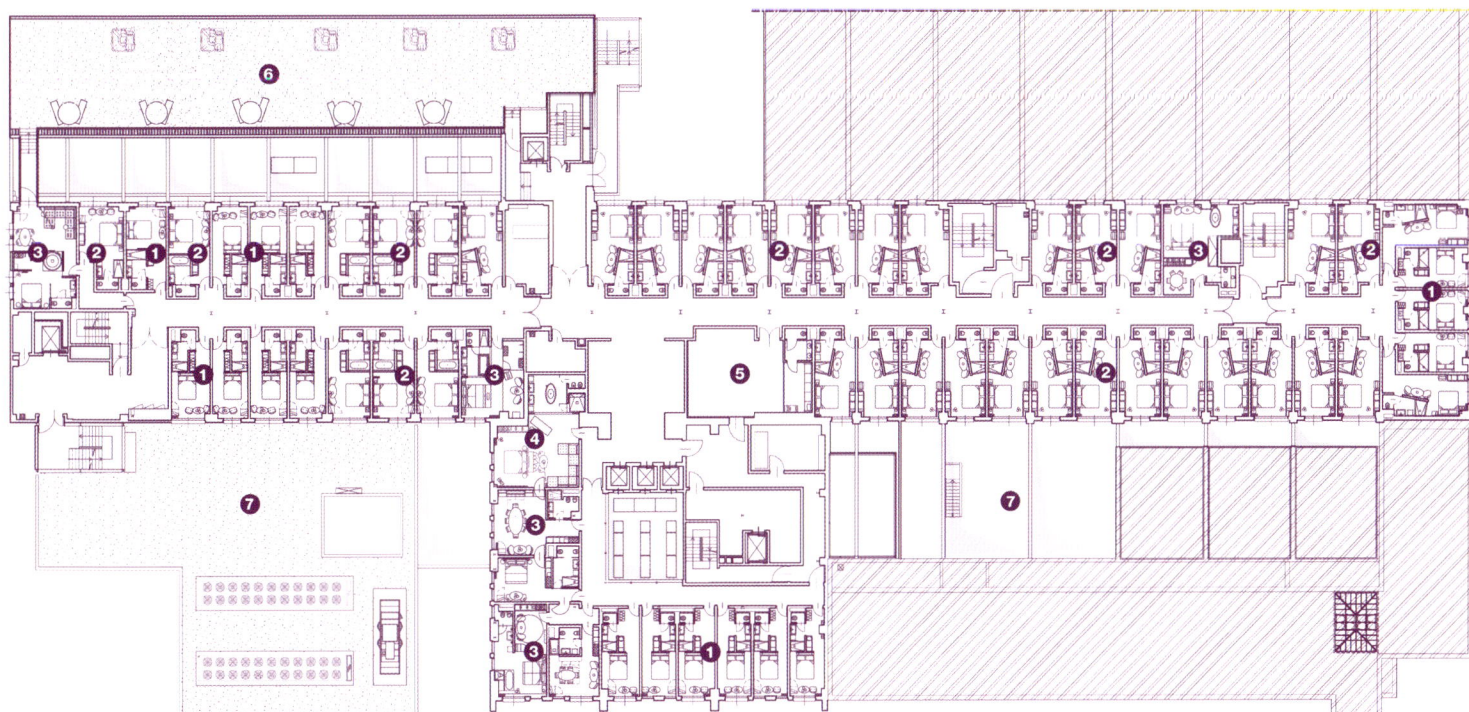

0 5 7 10m

↑↑标准间以实用为前提 / Standard room with utility as a precondition.
←浴室 / Bathroom.
↓标准客房平面图 / Standard room layout.

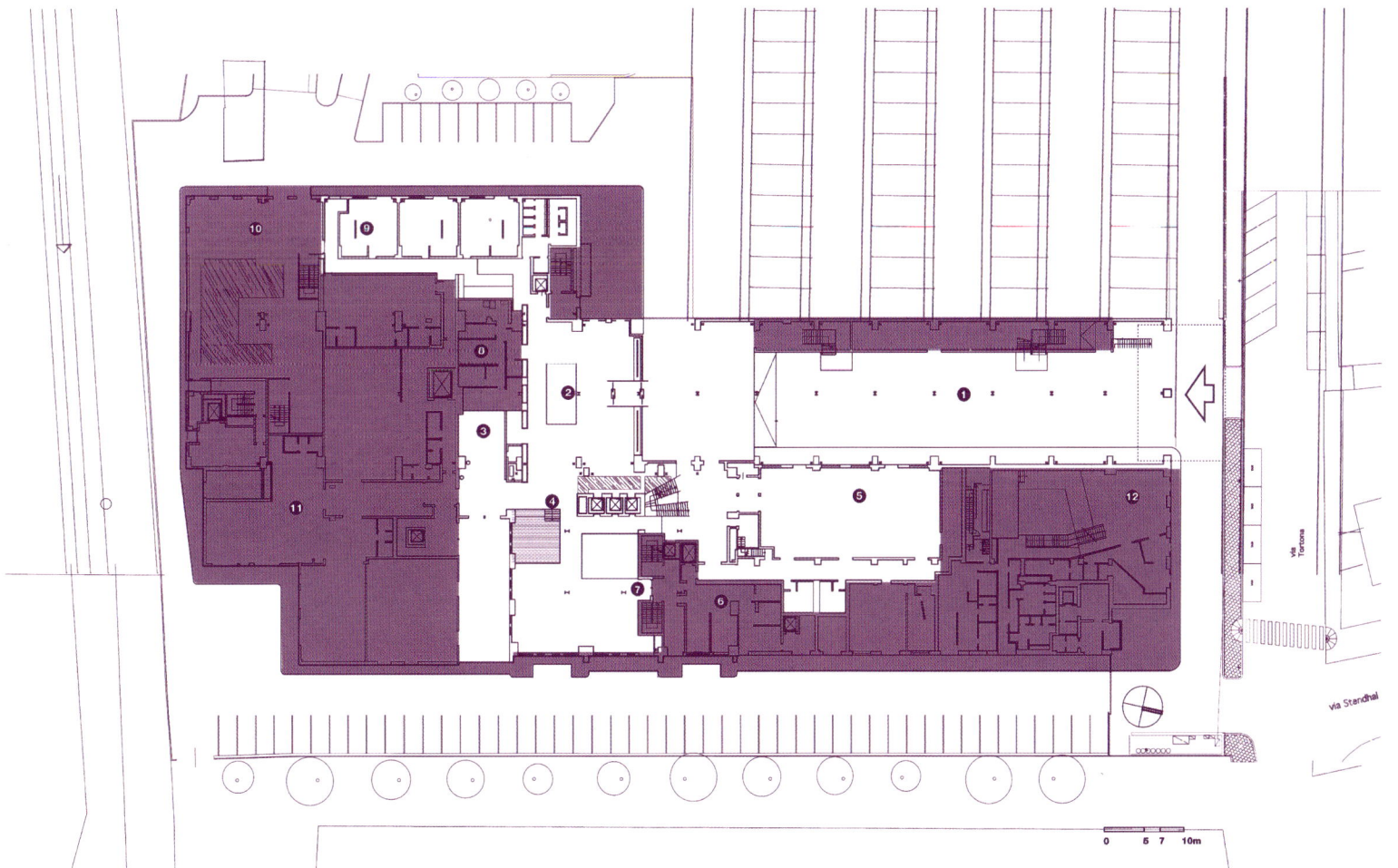

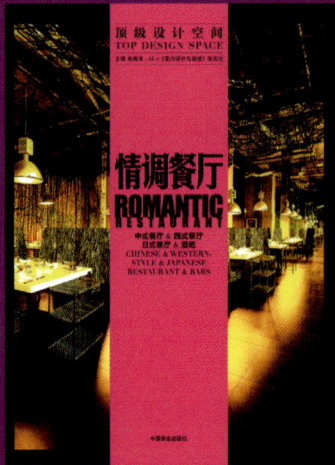
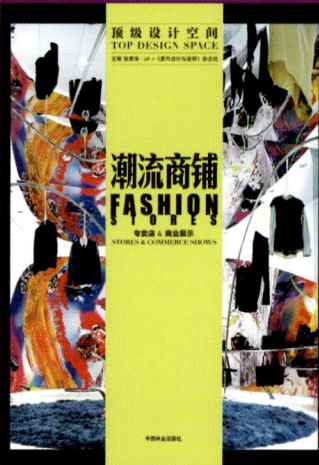
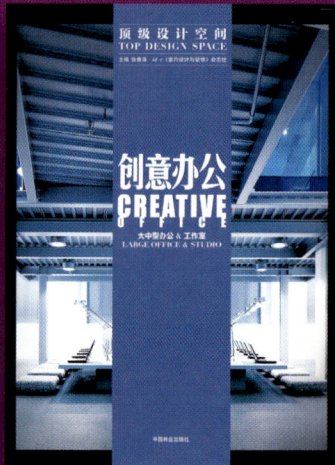
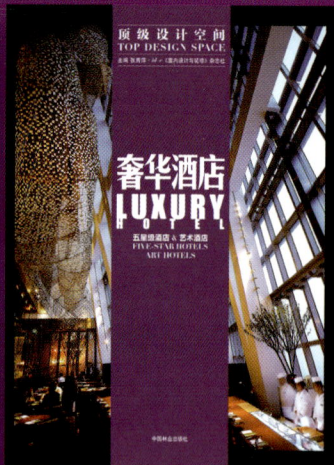

顶级设计空间 I——情调餐厅

ISBN 978-7-5038-5798-0
印装：四色精装
定价：248.00

顶级设计空间 II——纯粹商店

ISBN 978-7-5038-5797-3
印装：四色精装
定价：248.00

顶级设计空间 III——创意办公

ISBN 978-7-5038-5803-1
印装：四色精装
定价：248.00

顶级设计空间 IV——奢华酒店

ISBN 978-7-5038-5802-4
印装：四色精装
定价：248.00

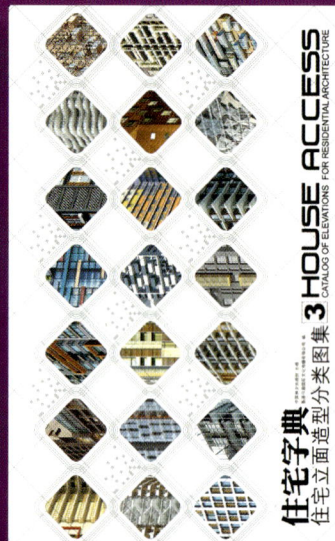

住宅字典
住宅立面造型分类图集. 1
ISBN 978-7-5038-5756-0

住宅字典
住宅立面造型分类图集. 2
ISBN 978-7-5038-5755-2

住宅字典
住宅立面造型分类图集. 3
ISBN 978-7-5038-5754-6

香港日瀚国际文化传播有限公司编
出版：中国林业出版社
印装：四色平装
开本：218mm×336mm
版次：2010年1月第1版
印次：2010年1月第1次
单册印张：20.25
单册定价：198.00

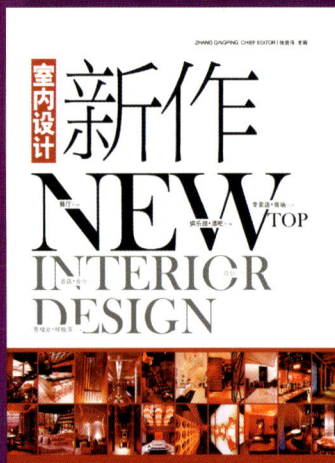

室内设计新作（上下卷）
张青萍 主编
ISBN 978-7-5038-5435-4
开本：230mm×300mm
页码：700
印装：软精装
定价：558.00元
出版时间：2009年6月

顶级样板房 I
张青萍 孔新民 主编
ISBN 978-7-5038-5709-6
开本：230mm×300mm
页码：360
印装：精装
定价：288.00元
出版时间：2009年10月

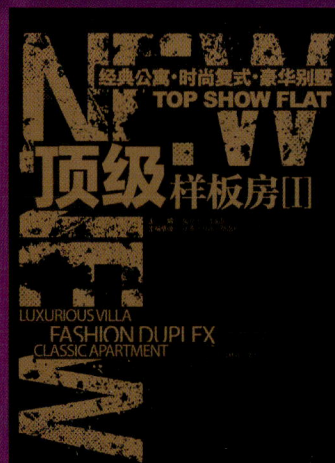

联系单位：中国林业出版社
地址：北京西城区德内大街刘海胡同7号
邮编：100009
销售客服：13641384559
出版客服：13810400238

网络支持：www.onetopspace.com